Letters of a Boy Soldier in the Civil War

Adelbert A. Buffum
Company B
24th Mass. Volunteer Infantry

edited with an introduction by Sarah Smith

max light books

A limited edition of this book was previously published for the Smith and Perry families, Christmas 2019.

For Mary Smith, who would have loved this,
and Kathy and Matt, who found the letters

Regimental flag of the 24th Massachusetts Volunteer Infantry
now in the State House, Boston

This is most of what we know about him, and even that isn't correct. He wasn't 18 when he enlisted; he was barely16:

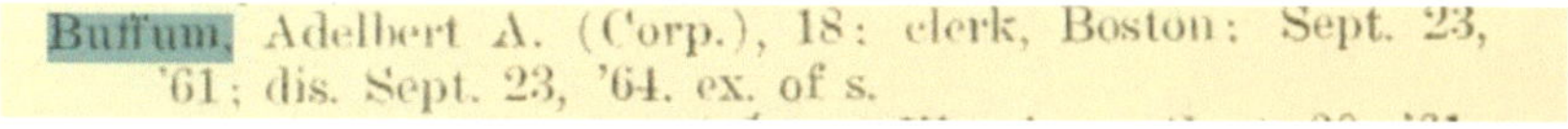
Buffum, Adelbert A. (Corp.), 18: clerk, Boston: Sept. 23, '61; dis. Sept. 23, '64. ex. of s.

His father, Carlon Edgar Buffum Sr., was born January 27, 1820, in Richmond Township, Cheshire, N.H., the son of James Buffum and Caroline Cook. Carlon Buffum moved to Boston, and on November 30, 1841, he married Adeline Abbott Boardman of Cambridge, Mass., daughter of Andrew Boardman and Adaline Hiscock.

In 1848 Carlon Buffum was named a clerk of the Boston Fire Department, probably not a full-time position. By 1862 he had a grocery store at 184 Friend St., Boston. Del had been working as a clerk there, probably from about the time he was 14. The store would provide various items to him in camp.

(This advertisement dates from 1872, when Carlon Sr. had moved to 191 Friend St.)

Carlon and Adeline Buffum's children were:

- Carlon Edgar jr., known as Edgar, born April 4, 1843, the "Dear Brother" of the letters
- Adelbert A., "Del" himself, born August 16 or August 18, 1845
- Theodore H. Dugan Kendrick, born February 1848; "Tede"

- Alonzo M., born August 10, 1850; "Loney"
- Addie Eveline, born Mar. 6, 1853
- William H.T., born February 6, 1856
- Charles Parker Ruggles, born May 16, 1858; "Charley"
- Mary E., born July 20, 1862; the "surprise baby" of the letters
- Flora May, born Jan. 14, 1866

So here they were at the beginning of the war: a prosperous grocer, his wife, and seven children, living in Melrose, with a grocery store near what's now North Station. In the fervor at the beginning of the war, Massachusetts was raising regiments, including the 24th Massachusetts Volunteer Infantry, which would be training at Readville (then part of Dedham, now part of Hyde Park).

Ordinarily, the oldest boy would have gone. That would have been Edgar. But as a child, Edgar had been dropped by his nurse; his back had never developed properly. He was extremely short-bodied, though very handsome (Longfellow called him "the handsomest man in Boston"). So he stayed a civilian, working for an indigo importer, and his younger brother went instead.

The handsomest man in Boston.

Alfred S. Roe's regimental history, *The Twenty-Fourth Regiment, Massachusetts Volunteers, 1861-1866* (Worcester MA: Twenty-Fourth Veterans Association, 1907), describes the war service of the regiment:

The 24th Regt. Mass. Vol. Inf., the New England Guards Regiment, was recruited at Camp Massasoit, Readville, under the personal supervision of Maj. Thomas G. Stevenson of the old New England Guards Battalion...who became its first colonel. Twenty-seven of the officers of the 24th Regt. came from this old battalion.

The recruits were mustered into the service from time to time as they arrived in camp, beginning early in September, 1861. The regiment remained at Readville until Dec. 9, when it left for Annapolis, Md., where it became a part of Foster's Brigade, Burnside's Coast Division. It sailed from Annapolis Jany. 9, 1862, as a part of the Burnside expedition bound for the coast of North Carolina. Here on Feb. 8, it was engaged with loss at Roanoke Island, and again March 14, it was in action at Newbern whom it suffered severely.

In May, 1862, divisions were formed, and the 24th became a part of Stevenson's (2d) Brigade, Foster's (1st) Division. About the 20th of March the regiment made an expedition to Little Washington, and a little later another up the Neuse River. Moving again toward Little Washington, on June 5 it met the enemy at Tranter's Creek, losing 5 killed and 9 wounded. Companies B and D formed a part of the garrison of Little Washington during the siege, July 8 to Sept. 12, and were engaged with loss Sept. 6. [Del Buffum was part of B Company.]

November 1, eight companies went on the Tarboro expedition and were engaged at Rawle's Mill, Nov. 2, with loss. On Nov. 12, Co. H was in action at Batchelder's Creek with loss.

...In the latter part of January, 1863, the regiment moved toward Charleston, S. C. It was in camp at St. Helena's Island, near Beaufort, S. C..... On Mar. 27 it was sent to Seabrook Island, Edisto Inlet, where it remained over three months. On July 10 it was transferred to James Island in front of Charleston, S. C. and remained on this and on Morris Island until Sept. 30, seeing little combat service.

Ordered to St. Augustine, Fla., it reached there Oct. 3, and remained through the rest of the fall and the winter following. Here 416 members of the regiment re-enlisted for three years. A part of the regiment was sent to Jacksonville, Fla., where it performed provost duty from Feb. 18 to April 24, 1864, when it was sent by transport to Gloucester Point near Yorktown, Va. Here the entire regiment was reunited about May 1....

It was engaged at Drewry's Bluff (also spelled Drury's Bluff), May 13-16, on the 16th having 11 men killed and 54 wounded or missing. It then moved to the north bank of the James, where it remained two months. On Aug. 14 to 16 it was heavily engaged at Deep Bottom, losing over 100 men of whom 20 were killed or mortally wounded....

The engagement near Deep Bottom, August 16, 1864, is commonly known in the regimental annals as Deep Run.

One of the companies of the 24th Mass. Volunteers is shown below.

Del Buffum's three-year enlistment was over in August 1864, From the letters, he'd been suffering from intestinal illness for as much as a year. He mustered out as a corporal and moved back to Boston.

He married twice. His first marriage, in 1866, to Mary E. Allen, produced two or three children.

- Adelbert Edgar was born in 1867. On December 22, 1891, he married Anne Maria Smith. In spite of having only an eighth-grade education, he became a lawyer in Cohasset, where he and "Annie" lived at 100 South Main St. He died in 1958 in Belmont, Mass..

- Marion Way Buffum was born in July 1869 and died a month later.

- A third child, Dora, is mentioned as being eight years old in the 1880 census.

After his first wife's death (or possibly divorce), he married for a second time in 1876. His second wife, Alice G. Foster, was born in 1851. They had one child, Lillian, who died of diphtheria in October 1881, at just under 3 years old.

In 1880 they were living with his father-in-law, Edward Foster, in Boston, and Del was working as a grocery clerk (maybe still at his father's store).

Del died of tuberculosis on September 4, 1884. Alice Foster Buffum outlived him by half a century and never remarried. In 1930 she was living in Newton, Massachusetts.

Of the rest of the family:

- Adeline Buffum died in Leominster in 1909.

- Carlon Edgar Buffum remained in business until 1882, when the store apparently went bankrupt. That may have had something to do with the marriage of *a* Carlon Edgar Buffum (same date and place of birth, same parents) to a Silvia Anderson Warner in 1880. Possibly he and Adeline had divorced. He died in 1896.

- Alonzo. M., "Loney", sadly, died on June 7, 1863. His final illness is mentioned in the letters.

- Mary E., "Nellie," died Oct. 12, 1886, at Lexington, at 24.

- Addie, William, and Charley grew up to have families.

- Flora May married George W. Elwell, a chauffeur, in 1890. They lived in New Haven. She died in 1941.

Theodore H. Dugan Kendrick, born February 1848, moved out West and became a cowboy. His niece, Sarah Buck, reported gleefully that the other cowboys were delighted by his four-barreled name. They called him "Theodore Dugan Kendrick Climax Sweepstakes Alphabet Buffum."

You have your choice of what the rest of his life was like. Genealogical sources report, prosaically, that he moved back East, lived in Newark where his niece and sister-in-law also were living, and died in 1921.

But for years I remembered a much more interesting story, which Sarah Buck told me as happening to him. T.H.D.K. had been sitting at the bar in a town out West when a man came in flourishing a revolver. "My name is Percy Akers," the man announced, "and I'm going to Hell." "Why?" asked T.H.D.K. Never ask why. The man shot him dead.

Of the two, I like the second story better. (And in spite of its tall-tale air, it actually happened, but to another relative, John O. French of Maxwell, Iowa, in 1887.)

Carlon Edgar Buffum jr., the "Dear Brother" of the letters, married Mary Foster Lawrence on September 2, 1880. (Delightfully, since he was almost a dwarf, their first child was a seven-months' baby; Edgar was born in Cohasset, Mass., on April 5, 1881.) He worked as a merchant and accountant, first in Massachusetts and then in New Jersey. He died December 27, 1914; he's buried in Fairmount Cemetery in Weymouth, Mass. Mary Foster Lawrence Buffum went to live with her daughter, Sarah, at 324 Myrtle St. in Manchester NH and died in 1938.

Edgar Buffum was one of the first motorcycle accident victims; though he survived, he was mentally compromised for the rest of his life and lived for many years at the Essex County Hospital for the Insane in Cedar Grove, N.J.

His younger sister, Sarah May Buffum, married Burton W. Buck in 1920. Their only child was Mary French Buck Smith, who had three children, Sarah, Kathryn, and Kendrick.

Back in the 1880s, Edgar Buffum had given his wife a desk. One day my sister Kathy and her son Matthew were poking around in the desk and found these letters…

And what you'll read here are the words of the letters: the Civil War as it was in one 16-year-old boy's eyes.

Sarah Smith
March 22, 2020

There are no known pictures of Adelbert Buffum.
This is 2nd Lieut. Frederick Rogers
of the 54th Massachusetts Volunteer Infantry.

Editorial Note

Adelbert Buffum's letters appear here just as they were written—both spelling and attitudes. He used the N-word freely. Like many Massachusetts soldiers, he resented officers leaving his regiment to serve in the Massachusetts 54th. He spoke casually of lynching, laughed at blackface, and helped pillage a town.

He was a person of his time.

In the world he helped make, I hope he would have thought differently.

The Letters

ROANOKE ISLAND, FEBRUARY 26, 1862

Page 1

Roanoke Island
Feb 26th 1861

Dear Brother

I received your letter on the 24th of this month I was very much pleased with the letters and papers on Monday Feb 24 after we came off drill we heard that there was a mail for us and there was cheering all over the camp and in a few minnits we heard the order fall in Co B and in a minnit there was a crowd around the orderly waiting for there letters there was about half a bushel of letters and papers for our company I am in good health and have been so ever since I left you say the sleighing is good there but I have not seen a flake of snow since I left Annapolis
I want you to write me as often

Roanoke Island Feb 26th [1862]

Dear Brother

I received your letter on the 24th of this month. I was very much pleased with the letters and papers on Monday Feb 24 after we came off drill we heard that there was a mail for us and there was cheereing all over the camp and in a few minnits we heard the order fall in Co B and in a minnit there was a croud around the orderly waiting for there letters thire was about half a bushel of letters and papers for our company I am in good health and have been so ever since I left you say the sleighing is good there but I have not seen a flake of snow since I left Annapolis I want you to write me as often

as you can you direct it to Co B
24th Regt Annapolis Burnsides
Cost Expedition and it will
come wherever we are and I would
like for you to send papers as often
as you can We had a hard time at
Hattrass but about as hard here before
we got settled the name of the
boat that the 24 was in is the Stemer
Oneida before we left Annapolis it
was called the Admeral we was
to have been the first Regt in the battle
but as the fleet was moveing towords
Roanoke our boat run aground and
the rest of the fleet passed us but the
next day small boats landed us and we
marched to the battle field through mud
up to our knees and water up to our
middle we got there just as the battery
was taken when General Burnsides
ordered us to follow up the retreting
rebels our Col asked if we should go
doubble quick no says the General

[2]as you can direct it to Co B 24th Regiment Annapolis Burnsides Co[a]st Expedition and it will come wherever we are and I would like you to send papers as often as you can. We had a hard time at Hattrass but about as hard here before we got settled the name of the boat that the 24 was in is the Steamer Guide before we left Annapolis it was called the Admeral we was to have been the first Regt in the battle but as the fleet was moveing twords Roanoke our boat run aground and the rest of the fleet passed us but the next day small boats landed us and we marched to the battle field through mud up to our knees and water up to our middle we got there just as the battery was taken when General Burnsides ordered us to follow up the retreting rebels our Col asked if we should go doubble quick no says the General

The unlikely name of the steamer was actually the Guide. It is referred to in the Letters of Col. William Bolton.

there is no use of tireing the men out
General Burnsides is one of the best men
that ever lived and I would follow
him wherever he may lead and I
guse every man in the Brigade would
well we followed the rebels up fast
~~as we were fresh~~ takeing prisoners as
we went untill we got almost into
the rebel camp when we halted to see
what was best to do it was not thought
safe to proceed father as our Regt. was
alone as the officers were talking the ma
mater over one of the rebels Col. came
rideing down with a flag of truce they
wanted hostilities stoped till morening
but our General said fight to night
or surrender so the rebel went back
to camp and we moved forward
Co. B was ahead so it was divided
in squads and sent ahead our Co.
alone took about 200 prisoners and in
about 2 hours more the rebels had
surrendered

[3] there is no use of tireing the men out General Burnsides is one of the best men that ever lived and I would follow him wherever he may lead and I guse every man in the Brigade would well we followed the rebels up fast ~~as we were fresh~~ takeing prisoners as we went untill we got almost into the rebel camp where we halted to see what was best to do it was not thought safe to proceed fa[r]ther as our Regt was alone as the officers were talking the mater over one of the rebel Cols came rideing down with a flag of truce they wanted hostilities stoped till morening but our General said fight tonight or surrender so the rebel went back to camp and we moved forward Co B was ahead so it was divided in squads and sent ahead our C° alone took about 200 prisoners and in about 3 [2?] hours more the rebels had surrendered

the last letter I wrote to mother had more particulars in so I will not write any more about the battle After we had been here about a week the prisoners were exchanged and the 24th marched them down to the wharf and they were put abord of one of our steamers and landed at Elisebarth City and we marched back to Camp and got some supper it was the only meal we had that day now we have drills the same as we had at Reidville about a week ago we had orders to pack up and go abord the boat so at 4 oclock in the morning we got up and packed up our knapsacks and took one days rations in our haversacks and canteen of water and Co C and B started for the battery where the battle was the march was abour 10 miles and we got there about 3 Oclock P.M and took 2 heavy cannon and put a roap on and the 2 Companys pulled them to one of the forts and marched back where we had stacked our guns when we heard that the order was contermanded so we marched back to camp and got there at dark the march that day was called about 25 miles and with knapsacks haversacks and canteens we were well

[4] the last letter I wrote to mother had more particulars in so I will not write any more about the battle. After we had been here about a week the prisoners were exchanged and the 24th marched them down to the wharf and they were put abord of one of our steamers and landed at Eliseberth City and we marched back to Camp and got some supper it was the only meal we had that day now we have drills the same as we had at Reidville about a week ago we had orders to pack up and go abord the boat so at 4 oclock in the mornnig we got up and packed up our knapsacks and took one days rations in our haversacks and canteen of water and Co E and B started for the battery where the battle was the march was about 10 miles and we got there about 3 Oclock PM and took 2 heavy cannon and put a roap on and the 2 Companys pulled them to one of the forts and marched back where we had stacked our guns when we heard that the order was contermandd so we marched back to camp and got there at dark the march that day was called about 25 miles and with knapsacks haversacks and canteens we were well

nigh worn out I never
sleapt so sound as I did
that night the next day
we was excused from drill
so we had nothing to do
but clean out guns and
now we are haveing good times here
the boys go out in the woods and
dig brier root and make pips I got
a piece and made a tip top pipe
but the worst of it is we cant get
but little tobacco and at a big price
and poor at that we can get bambo
reeds in the woods and the they make
good pipe stems the same as the
rebels had at fort Warren I wish
you would send me some
postage stamps in the next letter
that you write as I have not
got any and cannot get any here
and tell father to write to me

[5] nigh worn out I never sleapt so sound as I did that night the next day we was excused from drill so we had nothing to do but clean our guns and now we are haveing good times here the boys go out in the woods and dry brier root and make pip[e]s I got a piece and made a tip top pipe but the worst of it is we cant get but little tobacco and at a big price and poor at that we can get bambo reeds in the woods and ~~the~~ they make good pipe stems the same as the rebels had at fort Warren I wish you would send me some postage stamps in the next letter as I have not got any and cannot get any here and tell father to write to me

as there is nothing so pleasent
as to here from home when ever
there is a male there is cheering
all over the camp I would like
to get about a dozen betters evry
day and as many papers the bast
day that our Regt. was on gard
I was detailed to act as the Genls.
orderly and had to run all the forenoon
but in the afternoon I had nothing
to do but sit in the office (wich
is a little log house) and read and
smoak the next day the General
wanted me to come again because
I done so well and so I did and
those two days was the easiest
days I have spent scince I left
Annapolis we are a goin to be payd
off in a few days and then we
expect to start for some place
up the sound where we expect
to find some pretty hard fighting
but I do not know what place

[6]as there is nothing so pleasent as to here from home when ever there is a male there is cheering all over the camp I would like to get about a dozen letters evry day and as many papers the last day that our Regt was on gard I was detailed to act as the Gens orderly and had to run all the forenoon but in the afternoon I had nothing to do but sit in the office (wich is a little log house) and read and smoak the next day the General wanted me to come again because I done so well and so I did and them two days was the easiest days I have spent scince I left Annapolis we are a goin to be payd off in a few days and then we expect to start for some place up the sound where we expect to find some pretty hard fighting but I do not know what place

7

it is we heard news to day that it was reported that memphis Nashville and Savanna was taken by our troops I hope it is true I want to be in one stavin fight and then go home,
I guse I will clouse this letter now as I have not any more time now and the company is falling in for drill,

from your Brother
Dell,

P, S
Dont forget to sent me some Postage Stamps and write as soon and often as you can tell father to write as I have not had one letter from him yet,

from A. A. Buffum
for C. Edgar Buffum.

[7]it is we heard news to day that it was reported that memphis Nashville and Savanna was taken by our troops I hope it is true I want to be in one stavin fight and then go home, I guse I will closе this letter now as I have not any more time now and the company is falling in for drill

From your Brother

Dell.

P.S

Dont forget to sent me some Postage Stamps and write as soon and as often as you can. Tell father to write as I have not had one letter from him yet.

From A.A. Buffum

For C. Edgar Buffum.

[8][Endorsed in CEB's hand:

Adelbert A. Buffum

Roanoke Island

February 24/61 [sic]

CAMP LEE, N.C., MAY 15, 1862

May 15th /62
Camp Lee N.C.

Dear Brother

I recieved your letter last night and take this opertunity to answer it as it is a rainy day and we have not got to drill I would have wrote before but we have been moveing our camp quite often lately we have been out on pickit and was relieved by the 17th Regt. I saw All Crockett and George Leynde and John Grover on the Boat before they landed I heard that the 17th Regt. was about to land at Newbern so the Orderly and I with 2 or 3 men went aboard of the Boat and there we saw them as soon as there Regt. landed they marched out to camp and went on picket immediatly we are all well in our company we lost 1 in battle and one man

May 15th/62

Camp Lee N.C.

Dear Brother

I recieved your letter last night and take this opertunity to answer it as it is a rainy day and we have not got to drill I would have wrote before but we have been moveing our camp quite often lately we have been out on pickit and was relieved by the 17th Regt. I saw All Crockett and George Lynde and John Grover on the Boat before they landed I heard that the 17th Regt was about to land at Newbern so the Orderly and I with 2 orsmen went abord of the Boat and there we saw them as soon as there Regt landed they marched out to camp and went on picket immediately we are all well in our company we lost 1 in battle and one man shot himself at Fort Thompson and died of his wounds and one died since we came to this camp of the fever the rest of the company are all well you say it is not very hot there yet but it is hot enough we have to go in our shirt sleeves most all the time the grass is green and the trees are green and everything looks nice about the city of Newburn last Sunday our Regt went down to the city to meeting we stacted our guns in front of the meeting house and then marched into the house and the Band went up in the galery and played for the men to sing and the chaplin preached a sermon we had a very interesting meeting you say that Jim Leitch and Ed Holmes have got a stovepipe hat I should like to see Jim struting round with it on you say there are a good many letters on the way for me I have got

[2] shot himself at Fort Thompson and died of his wounds and one died csince we came to this camp of the fever the rest of the company are all well you say it is not very hot there yet but it is hot enough we have to go in our shirt sleves most all the time the grass is green and the trees are green and everything looks nice about the city of Newburn last Sunday our Regt went down to the city to meeting we stacted our guns in front of the meeting house and then marched into the house and the Band went up in the galery and played for the men to sing and the chaplin preached a sermon we had a very interesting meeting you say that Jim L[e?]itch and Ed Holmes have got a stovepipe hat I should like to see Jim struting round with it on you say there are a good many letters on the way for me I have got

them all and also the Box tell
father the Boots fit tiptop I would
have answered them all but have
not time I should like to see
Jonathan I guse I am haveing the
best time after all we have heard
bully news lately that Yorktown and
Norfork was evacuated and the
Merrimac was sunk the last we
heard was that McLellan was
following the retreating rebels I guse
they will get the divel before long
Fort Macon and Beaufort both are
taken 3 companies of our Regt have
gone to Washington N.C. to raise
a regt. of North Carolinans and
it is filling up fast our Col
is acting Brigidier General we were
payd off to day and I shall send
home a little money the drum
has beat for dinner so I must close
with my love to all in Haste
from your Brother A. A. Buffum

[3] them all and also the Box well father the Boots fit tiptop I would have answered them all but have not time I should like to see Jonathan I guse I am haveing the best time after all we have heard bully news lately that Yorktown and Norfolk was evacuated and the Merrimac was sunk the last we heard was that McLellan was following the retreting rebels I guse they will get the divel before long Fort Macon and Beaufort both are taken 3 companies of our Regt have gone to Washington N.C. to raise a regt of North Carolinans and it is filling up fast our Col is acting Brigidier General we were payd off to day and I shall send home a little money the drum has beat for dinner so I must close with my love to all in Haste

From your Brother A.A. Buffum

[4] [endorsed in CEB's handwriting]

Adelbert A. Buffum

Camp Lee N.C. May 15, 1862

WASHINGTON, N.C., JULY 22, 1862

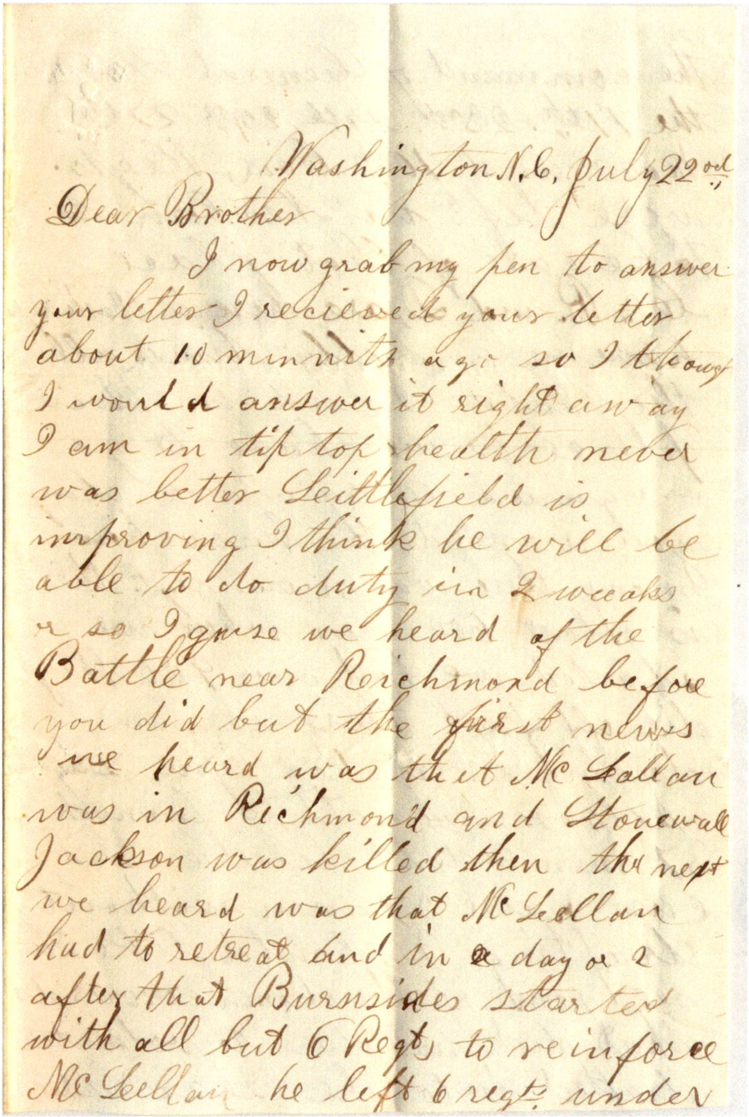

Washington N.C. July 22nd

Dear Brother

I now grab my pen to answer your letter I recieved your letter about 10 minnits ago so I thought I would answer it right away I am in tip top health never was better Littlefield is improving I think he will be able to do duty in 2 weeks or so I guse we heard of the Battle near Richmond before you did but the first news we heard was that Mc Lellan was in Richmond and Stonewall Jackson was killed then the next we heard was that Mc Lellan had to retreat and in a day or 2 after that Burnsides started with all but 6 Regts to reinforce Mc Lellan he left 6 regts under

Washington, N.C., July 22nd

Dear Brother

I now grab my pen to answer your letter I recieved your letter about 10 minnits ago so I thougt I would answer it right away I am in tip top health never was better Littlefield is improving I think he will be able to do duty in 2 weeaks or so I guse we heard of the Battle near Richmond before you did but the first news we heard was that McLallan was in Richmond and Stonewall Jackson was killed then the next we heard was that McLellan had to retreat and in a day or 2 after that Burnsides started with all but 6 Regts to reinforce McLellan he left 6 regts under

the command of General Foster
the 17th. 23rd. 24th. 25th 27th
Mass and 10th Con, Regts
were left in Newbern to
take care of that place
Co. B. and D, are here in Washington
haveing a bully time all
the fault I find in this
place is we have to go
on guard every other day
you wanted to know how
many there was in our regt there
is about 600 men when we
started we had over 900 they
die off fast there is some
dyeing every day you say
nelsore has gott to furnish
37 men I don't believe they
can raise 37 men without
drafting I should like to
see some of the damn cowards
that did not dare to volanter
drafted into the service

[2] the command of General Foster the 17th, 23rd, 24th, 25th, 27th Mass and 10th Con[n] Regts were left in Newbern to take care of that place Co B and D, are here in Washington haveing a bully time all the fault I find in this place is we have to go on guard every other day you wanted to know how meny there was in our regt there is about 600 men when we started we had over 900 they die off fast there is some dyeing every day you say Melrose has got to furnish 37 men I dont believe they can raise 37 men without drafting I should like to see some of the d—n cowards that did not dare to volunter drafted into the service

you say that you heard good
playing on Gilmores band
but we do not hear very good now
as 3 of the Band has died and
Gilmore has gone to Boston on
a furlough you say Jonathan
is wounded. I would give 5 $
to see him I vow I would
I was talking with one of our
boys the other day and I asked
him where he used to work he
said he worked for P B Holmes
and the next time you see
Ed Holmes you ask him if
he remembers a fellow that
used to work for his father by
the name of Harry Bent
Bent said he used to see me
most every day going by the shop
I find somebody that I used to
know round home quite often
it is a right smart hot day
and as I just come off guard

[3] you say that you heard good playing in Gilmores band but we do not hear very good now as 3 of the Band has died and Gilmore has gone to Boston on a furlough you say Jonathan is wounded I would give 5 $ to see him I vow I would I was talking with one of our boys the other day and I asked him where he used to work he said he worked for P B Holmes and the next time you see Ed Holmes you ask him if he remembers a fellow that used to work for his father by the nam of Harry Bent Bent said he used to see me most evry day going by the shop I find sombody that I used to know round home quite often it is a right smart hot day and as I gust come off guard

this morning I must
lay down and get a little
sleep so I will close
with my love to all
from your Brother
Dell

P. S. you say you will
send me papers evry weak
so I will send you a paper
every weak this paper is
one that our Boys prints one
man from Co. B and one from
Co. D it is printed here in
Washington

from A. A. Buffum
Co. B, 24th Regt Mass Vol

Adelbert A. Buffum
Washington
July 5. 1863

[4] this morning I must ly down and get a little sleep so I will close with my love to all from your Brother

Dell

P.S. you say you will send me papers evry weeak so I will send you a paper every weeak this paper is one that our Boys prints one [man?] from Co B and one from Co D it is printed here in Washington

From A.A. Buffum

Co B, 24th Regt Mass Voll

[Endorsed, upside down, in CEB's handwriting

Adelbert A. Buffum

Washington, N.C. July 22, 1862]

WASHINGTON, N.C., AUGUST 8, 1862

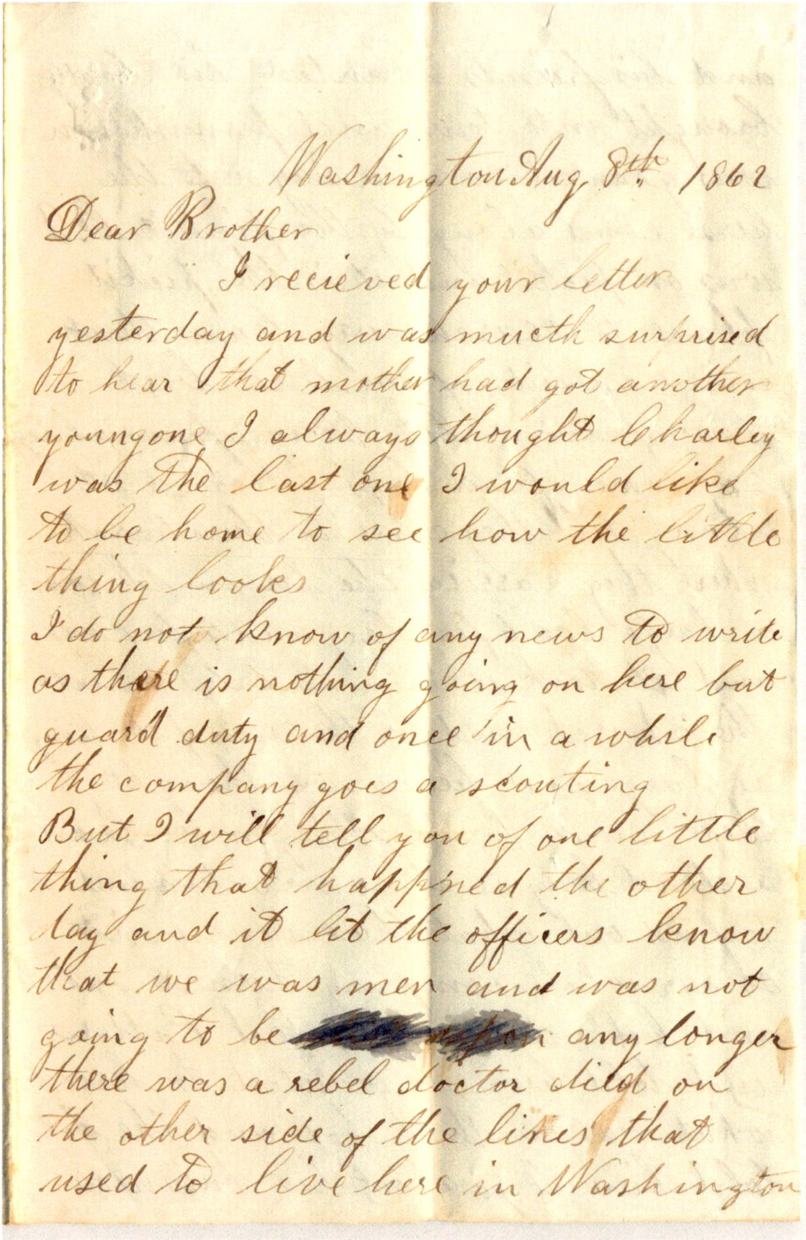

Washington Aug 8th 1862

Dear Brother

I recieved your letter yesterday and was much surprised to hear that mother had got another youngone I always thought Charley was the last one I would like to be home to see how the little thing looks

I do not know of any news to write as there is nothing going on here but guard duty and once in a while the company goes a scouting

But I will tell you of one little thing that happened the other day and it let the officers know that we was men and was not going to be [redacted] any longer there was a rebel doctor died on the other side of the lines that used to live here in Washington

Washington, Aug 8th, 1862

Dear Brother

I recieved your letter yesterday and was mucth surprised to hear that mother had got another youngone I always thought Charley was the last one I would like to be home to see how the little thing looks

I do not know of any news to write as there is nothing going on here but guard duty and once in a while the company goes a scouting

But I will tell you of one little thing that happned the other day and it let the officers know that we was men and was not going to be ["shit upon" effaced in heavy pencil] any longer there was a rebel doctor died on the other side of the lines that used to live here in Washington

and his friends wanted his body
brought into town and permishion
was granted so in a day or 2 the
herse came along by the pickit Co. D
was on pickit and it is the pickit
place to search evry team that
comes inside of our lines but
these fellows let them pass along
so some of our boys followed it
along till they came to the church
where they carried the box in it was
a large box 6 feet long and 4 feet wide
and it took 12 men to take in into
the church and for the reson of its being
so large and heavy we thought
there must be something more than
a dead body in it so we determinded
to find out what thare was in it
so our whole company turned out
and Cobb went up to Col Potters
and told him we wanted to know
what was in that box and the Col
told him to go about his buisness

[2] and his friends wanted his body brought into town and permishtion was granted so in a day or 2 the herse came along by the pickit Co D was on pickit and it is the pickit place to search evry team that comes inside of our lines but these fellows let them pass along so some of our boys followed it along till they came to the church where they carried the box in it was a large box 6 feet long and 4 feet wide and it took 12 men to take in into the church and for the reson of its being so large and heavy we thought there must be something more than a dead body in it so we determinded to find out what there was in it so our whole company turned out and Cobb went up to Col Potters and told him we wanted to know what was in that box and the Col told him to go about his buisness

what hat a private to do with any
shuch thing he called us a pack of
damn cowards then Cobb told him
if he did not have it opened we
would and he could not help himself
so we all went up to the Church
and the officers had it opened but
it was all right there was a body
in a coffin with pitch and stuff
around it and another rough box
out side of that be we did not
know but the box was full of
pistols or guns and as there is plenty
of citesens that are rebels to the
back bone and there was 2 of our
pickits shot once by a citisen
I do not know of any more to
write now sot I will close
with my love to all
from your Brother Del

PS I will send you a paper with
this letter

[3] what hat a private to do with any shuch thing he called us a pack of d—n cowards then Cobb told him if he did not have it opened we would and he could not help himself so we all went up to the Church and the officrs had it opened but it was all right there was a body in the coffin with pitch and stuff around it and another rough box outside of that be we did not know but the box was full of pistols or guns and as there is plenty of citisens that are rebels to the backbone and there was 2 of our pickits shot once by a citisen I do not know of any more to write now so I will close with my love to all

from your Brother Del

P S I will send you a paper with this letter

[4] [Endorsed in CEB's handwriting:

Adelbert A. Buffum

Washington N.C. Aug. 8 1862]

Washington, N.C., Sept. 1, 1862

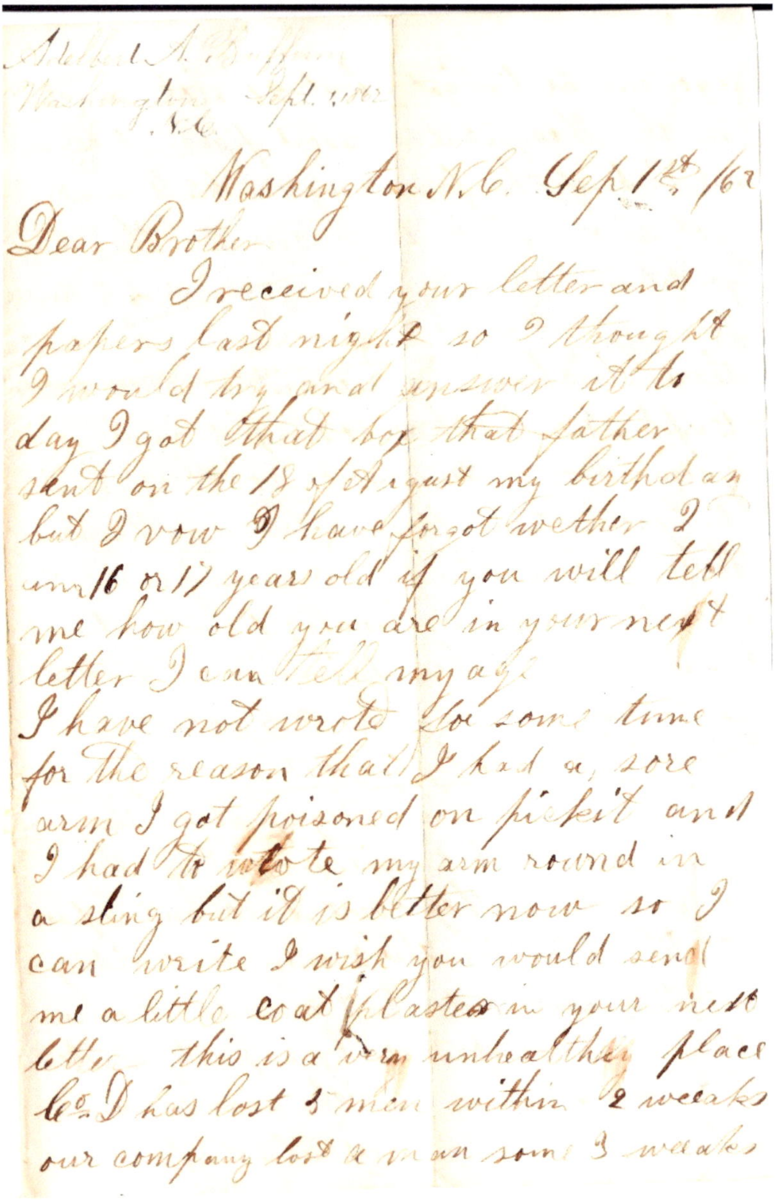

Adelbert A. Buffum
Washington Sept. 1862
N.C.

Washington N.C. Sep 1st /62

Dear Brother

I received your letter and papers last night so I thought I would try and answer it to day I got that box that father sent on the 18 of August my birthday but I vow I have forgot wether I am 16 or 17 years old if you will tell me how old you are in your next letter I can tell my age

I have not wrote for some time for the reason that I had a sore arm I got poisoned on pickit and I had to wrote my arm round in a sling but it is better now so I can write I wish you would send me a little coat plaster in your next letter this is a very unhealthy place Co. D has lost 3 men within 2 weeks our company lost a man some 3 weeks

[Endorsed, Adelbert A. Buffum

Washington N.C. Sept 1, 1862]

Washington N.C. Sep 1st/62

Dear Brother

I received your letter and papers last night so I thought I would try and answer it to day I got that box that father sent on the 18 of August my birthday but I vow I have forgot wether I am 16 or 17 years old if you will tell me how old you are in your next letter I can tell my age

I have not wrote for some time for the reason that I had a sore arm I got poisoned on pickit and I had to [tote?] my arm round in a sling but it is better now so I can write I wish you would send me a little coat [court] plaster in your next letter this is a very unhealthy place Co D has lost 5 men within 2 weeaks our company lost a man some 3 weeaks

ago our Co. has got about 21 sick men
in the Hospital a good part of them
are sick of the fever & ague Co. I has
got more sick than our Co. they have
got 30 men for duty and we have 45 men
for duty out of 93 3 of our men have got
there discharge on acount of there health
Capt. Austin has resinged and is
going home tomorrow so there will
be promotions through the whole company
there is 40 recrutes for our 2 companies at
Newbern 3 of our Cavalry 1 Corporal and
2 men were taken prisoners on pickit
yesterday morning about 4 oclock
there is 3 Cavalry pickit on every road
where our pickit is they patrol the
road out about 5 miles and if they
see enything they come back and report
it to us while the 3 men were riding
out they were out about 2 miles from us
they were surprised by a party of rebels
that were laying in the woods and
when the Cavalry were abrest of them

[2]ago our Co has got about 21 sick men in the Hospital a good part of them are sick of the fever & ague Co D has got more sick than our Co they have got [30?] men for duty and we have 45 men for duty out of 93 3 of our men have got there discharge on acount of there health Capt Austin has resinged and is goign home tomorrow so there will be promotions through the wholl company there is 40 recutes for our 2 companies at Newbern 3 of our Cavalry 1 Corporal and 2 men were taken prisoners on pickit yesterday morning about 4 oclock there is 3 Cavalry pickit on every road where our pickit is they patrol the road out about 5 miles and if they see anything they come back and report it to us while the 3 men were riding out they were out about 2 miles from us they were surprised by a party of rebels that were laying in the woods and when the Cavalry were abrest of them

they jumpt out into the road and
caught the horses by the bridles and
pointed pistols at that there heads
and called for them to surrender one
of the Cavalry drew his saber and
swaved it at a nigger and told him
to go in and report it that had to
surrender as there was 50 rebels to
there 3 the nigger came running in
and reported it and the whole Co
of Cavelry was in the saddle and
flying over the road in about 5 minuts
they went out about 6 miles but
could not find them it is believed
that they will be killed because
the rebels have had orders to spare
no Cavalry nor mareines the rebels
caught 2 of our Cavelry before the
same way the other day one of the
N.C. Companies went out scouting and
caught 11 rebels and they are trying to
get square there is some of us out
scouting most every day and I dont

[3] they jumpt out into the road and caught the horses by the bridles and pointed pistols at that there heads and called for them to surrender one of the Cavalry drew his saber and waved it at a nigger and told him to go in and report it that had to surrender as there was 50 rebels to there 3 the nigger came running in and reported it and the whole Co of Cavelry was in the saddle and flying over the road in about 4 minutes they went out about 6 miles but could not find them it is believed that they will be killed because the rebels have had orders to spare no Cavalry nor mareines the rebels caught 2 of our Cavelry before the same way the other day one of the N.C. Companies went out scouting and caught 11 rebels and they are trying to get square there is some of us out scouting most every day and I don't

see how in the divel the rebels got down there without us knowing it

I dont know of any more news to write non but I will send you a paper and a half the reason they could not git a whole one out is because one of the printers had to go on duty and the other one he that belonged to our Co. is dead

I cant think of anything more to write now so I will close with my love to all

in haste from your Brother

Dell

if it had not been for my sore arm I would have wrote sooner. wrote sooner

wrote sooner.

wrote sooner.

wrote sooner.

Wrote sooner.

wrote sooner.

wrote sooner.

wrote sooner.

wrote sooner.

wrote sooner.

Wrote Sooner.

Dell

[4]see how in the divel the rebels got down there without us knowing it I dont know of any more news to write no[w] but I will send you a paper and a half the reason they could not git a whole one out is because one of the printers had to go on duty and the other one he that belonged to our Co is dead.

I cant think of anything more to write now so I will close with my love to all.

in haste from your Brother

Del

if it had not been for my sore arm I
would have wrote sooner, wrote sooner
wrote sooner,
wrote sooner,
wrote sooner,
wrote sooner,
wrote sooner,
wrote sooner,
wrote sooner,
wrote sooner,
wrote sooner,
Wrote Sooner.
Dell

Pickit Camp, December 7, 1862

Corporal A. A. Buffum,

Ans. Jany 5/63. 8 Jan 13" 5" to 13"

Picked Camp Dec 7th/62

Dear Brother

I now take my pen in my clutches to write you a few lines I will now try and give you some account of the expedition. on the 28th of last Oct we had orders to clean up our guns and equipments and get ready for review on the morrow and at supper time we was told we was to go on a march and to be ready by day light the next morning and by 7 oclock the next morning we was on board of the boat in light marching order wee went to Washington and stayed there 3 days waiting for the rest of the force and on the 1st of Nov we started on the march the 1st Brigade under the command of Col Stevenson the 2nd under the command of Col Amory Col of the 17th Regt. the 3rd by the Col of the 10th Ct

[Endorsed by CEB, "Arr. Jan 5/63 and Jan 13th 5th to 13th"]

Pickit Camp, Dec 7th/62

Dear Brother,

I now take my pen in my clutches to write you a few lines I will now try and give you some account of the expedition on the 28th of last Oct we had orders to clean up our guns and equipments and get ready for review on the morrow and at supper time we was told we was to go on a march and to be ready by day light the next morning and by 7 oclock the next morning we was on board of the boat in light marching order we went to Washington and stayed there 3 days waiting for the rest of the force and on the 1st of Nov we started on the march the 1st Brigade under the command of Col Stevenson the 2nd under the command of Col Amory Col of the 17th Regt the 3rd by the Col of the 10th Ct

we started from Washington and
marched all that day sometimes
we had to take off our shoes and
stockings and roll up our pants and
march through water up to our knees
for a mile or so at about 6 Oclock in
the we had a fight witch
lasted about 2 hours with the loss of
4 killed and a number wounded
there was 1 killed in Co. E of our regt
2 out of the 44th and 1 marien it
was mostly artillery fighting it ended
in the retreat of the rebels and we
followed them till 2 oclock the next
morning when we halted till
day light the next morning makeing
the march that day 20 miles and
I tell you we was all darn near
played out when we halted as soon
as we got a little coffee we started
and marched till about 3 Oclock when
we came to Williamston the rebels
retreating as fast as we advanced

[2] we started from Washington and marched all that day sometimes we had to take off our shoes and stockings and roll up our pants and march through water up to our knees for a mile or so at about 6 Oclock in the [word omitted] we had a fright [fight] witch lasted about 2 hours with the loss of 4 killed and a number wounded that was 1 killed in Co. E of our regt 2 out of the 44th and 1 marein it was mostly artillery fighting it ended in the retreat of the rebels and we followed them till 2 oclock the next morning when we halted till day light the next morning making the march that day 20 miles and I tell you we was all darn near played out when we halted as soon as we got a little coffee we started and marched till about 3 Oclock when we came to Williamston the rebels retreating as fast as we advanced

and I tell you it was gay sport
when we got in the city the
inhabitancs all left somewhere
I dont know where and left there
houses and stores just as there were
and the Boys smashed into the
stores and got any quanty of tobacco
and everything you could think of
they would go in and take silk
cloth and all kinds of goods and
chuked them into the street the Boys
could not carry anything so they
distroyed them ~~a~~ some of our
fellers went into a house and stove
up 3 nice pianfortes and throughed
them into the street they raised
the very divel there was plenty of
whiskey round and a good many
were geting tipsy so at about 5 oclock
we ordered to fall in it took some time
to get the men together but after a
while we got started and marched
till dark when we come to a place

[3] and I tell you it was gay sport when we got in the city the inhabitancs all left somewhere I dont know where and left there houses and stores just as there were and the Boys smashed into the stores and got any quanty of tobacco and everything you could think of they would go in and take silk cloth and all kinds of goods and chuked them into the street the Boys could not carry anything so they distroyed them some of our fellers went into a house and stove up 3 nice pianofortes and throughe them into the street they raised the very divel there was plenty of whiskey round and a good many were getting tipsy so at about 5 oclock we ordered to fallin it took some time to get the men together but after a while we got started and marched till dark when we come to a place

called Hamilton but we was not
allowed in the city because they
raised the dibel so in Willamston
so we camped out side the city till morning
when we started and marched till we
could hear the cars at Tarborra and
for some cause or other we turned back
for home and got there on the 11 of Nov
all tired out some of the boys were
barefooted some in there stocking feet
all hands of us had sore feet
after we got our supper we all turned in
to get a good sleep something we had
not seen for a fortnight but we had not
been layed down but a half hour when
we had a long roll and we fell in and
was ordered on picket as we was about
the only regt. that had got in out picket
had been ataced on 2 roads and driven in
on one of then so our regt was put on and
stayed till morning at the same
time Co. H of our regt. was attaced
and 2 men killed so after geting
one nights sleep we was sent up
to reinforce Co. H and have been
here ever since. I must close this
letter as I have not got any more
paper from your Brother
Please write soon A A Buffum

[4] called Hamilton but we was not allowed in the city because they raised the divel so in Williamston we camped outside the city till morning when we started and marched till we could hear the cars at Yarboroa and for some cause or other we turned back for home and got there on the 11 of Nov barefooted some in there stocking feet all hands of us had sore feet after we got our supper we all turned in to get a good sleep something we had not seen for a fortnight but we had not layed down by a half hour when we had a long roll and we fell in and was ordered on picket as we was about the only regt that had got in out picket had been ataced on 2 roads and driven in one one of them so our regt was put on and stayed till morning at the same time Co H of our regt was attaced and 2 men killed so after geting one nights sleep we was sent up to reinforce Co H and have been here ever since. I must close this letter as I have not got any more paper from your Brother Please write soon A A Buffum

OFF BEAUFORT, S.C., JAN 29TH/ 1863

Off Beaufort Jan 29th 1863

Dear Brother

I recieved your letter of the 15th yesterday with Charles picture the little buger has grown like thunder [illegible] I have been expecting mothers and the babys picture for some time I hope she will send it in the next letter I am in tip top health now I weigh more than ever I did in my life I weigh 148 lbs all the Melrose Boys at Newbern are all well Serg Leake has been promoted to 1st Lieut of one of the N.C. Companies at Washington but has been detached as Civil Engineer he has got Washington fortified tip top he has got a big ditch all round the city I saw Al Crockett the other day and he said George had been promoted again he is a Sergent now [illegible] says George fights like the devil he dont care for the rebels

Dear Brother

I recieved your letter of the 5th yesterday with Charleys picture the little buger has grone like thunder aint he I have been expecting mothers and the babys picture for some time I hope she will send it in the next letter I am in tip top health now I weigh more than ever I did in my life I weight 148 lbs all the melrose Boys at Newbern are all well Sergt Lake [Leake?] has been promoted to 1st Liut of one of the N.C. Companies at Washington but has been detached as Civil Engineer he has got Washington fortified tip top he has got a big ditch all round the city I saw All Crockett the other day and he said George had been promoted again he is a Sergent now he says George fights like the divel he dont care for the rebels

tell Father that I got both of the boxes
and they were in first rate order tell
him I like those shirts bully
our Regt and the 10th Conn. are on board of boats
laying off Beaufort 4 Companies of us are on
board of the Scanner Highlander the rest of the
Regt are on the Steamer Guide there is a
very large fleet here they say we are bound
for Wilmington N.C. and Charleston
S. C. they say there is a large fleet
going to join us at Port Royal our
fellers dont expect to go back to
Newbern again there is none of the
Melrose boys with us as there is only
2 Regts of our Brigade with us General
Stevenson is with us the 44th Mass
and 5th R. I. belong to our Brigade
but are left at Newbern [illegible] is
got out of the Hospital and is quite
smart again he is at Newbern for now
all the 9 months troops are left at
Newbern we expect to see some

[2] tell father that I got both of the boxes and they were in first rate order tell him I like thoes shirts bully

Our Regt and the 10 Conn are on board of boats laying of Beaufort 4 Companies of us are on board of the Sconner [schooner] Highlander the rest of the Regt are on the Steamer Guide there is a very large fleet here they say we are bound for Willmington N.C. And Charleston S.C. They say there is a large fleet going to join us at Port Royal our fellers don't expect to go back to Newbern again there is none of the Melrose boys with us as there is only 1 Regt of our Brigade with us General Stevenson is with us the 44th mass and 5th R.I. Belong to our Brigade but are left at Newborn Littlefield is got out of the Hospital and is quite smart again he is at Newbern for now all the 9 months troops are left at Newbern we expect to see some

tough fighting in this Expedition
our Regt was the only regt of our Brigade
that come and Foster was going to putt
us in a new Brigade and Stevenson
told him if he done that he would
take off the stars and come back as
Col. of the [illegible] so they sent the
10 Conn. here and Stevenson took us as
a Brigade, I dont know as you can
read this letter as it is a very poor
place I have to write our Co. is down
in the hole I am sitting in my bunk
and writeing this by the dim lite of
a lantern and I cant see wether
my pencel is on the line or not
the Capt. has just come down and [illegible]
sayes the mail goes in an hour and
as I dont know of any thing more
to write now I will close with
my love to all the next time I will
try to write a longer letter
from your Brother Adelbert
to
Edgar

[3] tough fighting on this Expedition our Regt was the only one of our Brigade that come and Foster was going to put us in a new Brigade and Stevenson told him if he done that he would take off the stars and come back as Col of the 24th Regt so they sent the 10 Conn here and Stevenson took us as a Brigade. I dont know as you can read this letter as it is a very poor place I have to write our Co is down in the hole I am siting in my bunk and writeing this by the dim lite of a lantern and I can't see wether my pencil is on the line or not the Capt has just come down and says the mail goes in an hour and as I dont know of anything more to write now I will close with my love to all the next time I will try to write a longer letter

from your Brother Adelbert

to Edgar Buffum

[4]

Endorsed in CEB's handwriting

Adelbert A. Buffum

Off Beaufort N.C. Jany 29 1863

(Schr. Highlander)

St. Helena Island, S.C., March 15, 1863

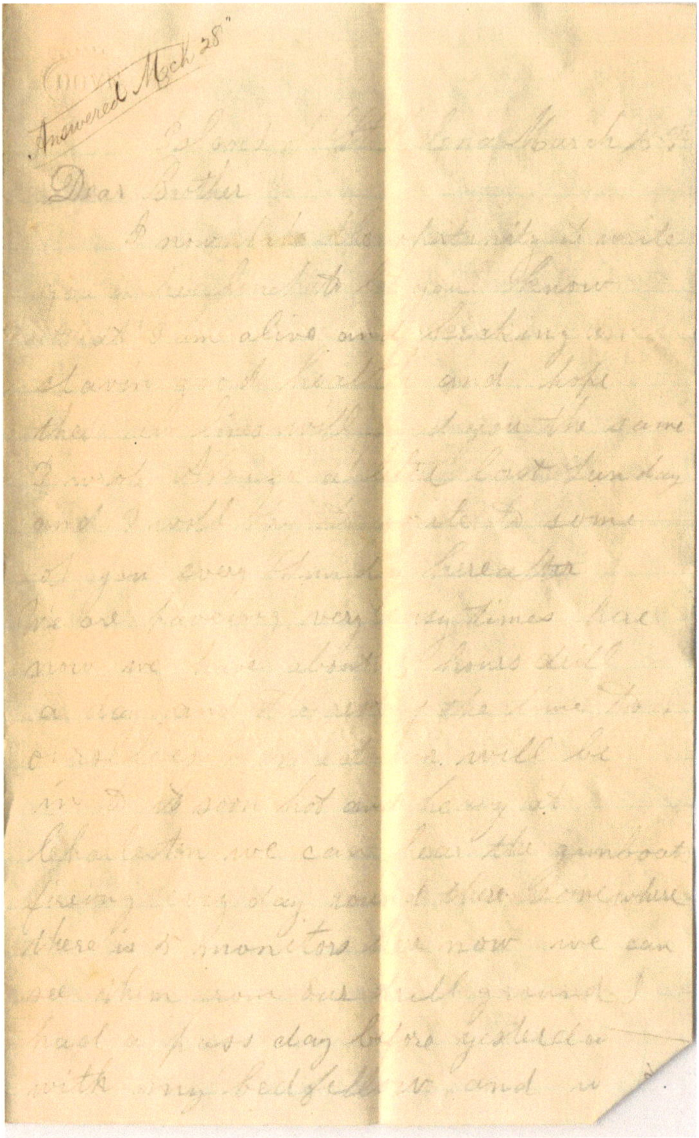
Answered Mch 28"

[illegible] March 15th

Dear Brother

I [illegible] write
you [illegible] know
that I am alive and [illegible]
[illegible] health and hope
[illegible] the same
I [illegible] last Sunday
and I will [illegible] to some
of you every [illegible] hereafter
We are having very [illegible] times here
now [illegible] about [illegible]
[illegible]
[illegible] will be
in [illegible] at
Charleston we can hear the gunboat
firing [illegible] day [illegible]
there is 5 monitors [illegible] now we can
see them [illegible] I
had a pass day before yesterday
with my bedfellow and [illegible]

[Written in pencil, which is rubbed and difficult to read.

Endorsed in CEB's hand, Answered Mch 28]

Island of St Helena March 15th [1863]

Dear Brother

I now take the opertunity to write you a few lines to let you know that I am alive and kicking in stavin good health and hope these few lines will find you the same I wrote Alonzo a letter last Sunday and I will try to write to some of you every Sunday herafter

We are haveing very easy times here now we have about 3 hours drill a day and the rest of the time to ourselves I expect we will be in to some hot and heavy at Charleston we can hear the gunboats fireing every day round there somewhere there is 5 monitors here now we can see them from our drill ground I had a pass day before yesterday with my bedfellow and we

went on Board of one of them and
Stavin smart things they are too
when we went aboard the hands were
busy in shushing down the deck and
current with grea about a quarter of
and inch deep we went in the current
and see the big gun that fires 35 lbs
of powder at one charge the port holes
are fixed so when they shove the
muzel of the gun out it fills the port hole
all up and when they fire the gun
jumps back and the Port hole close
quicker than [illegible] they are great
things I tell you it would be imposible
to board one of them there is little holes
fixed in the decks so they can
shove shells up and the minit
they strike the deck they explode
I was aboard the one that was up
to Fort Mcallister and Sunk the
Nashville they said they run up
within fireing distance and

[2] went on Board of one of them and Stavin smart things they are too when we went aboard the hands were buisy in shushing down the deck and turrent with greese about a quarter of and inch deep we went in the turrent and see the big gun that fires 35 lbs of powder at one charge the port holes are fixed so when they shove the muzel of the gun out it fills the Porthole all up and then they fire the gun jumps back and the Port hole closes quicker than a wink they are great things I tell you it would be imposible to board one of them there is little holes fixed in the deck so they can shove shells up and the minit they strike the deck they explode I was aboard the ones that was up to Fort Mcallister and sunk the Nashville they said they run up within fireing distance and

dropped anchor and piped away and when they piped all hands down to dinner they all went down and let the Fort fire away she got struck a good many times but it did not hurt her any only made a few dents in her we all expect to have a darn hard fight when we do get at Charleston the Pickets were attack over on the other side that is Hilton Head last night and one of the Cavalry was killed I believe

I got your last papers of the 4 of March and Wednesday they was the only ones for the Company there was quite a lot of news they had to go the rounds through the Company I see they got the news in about General Stevenson and the cheap troops [illegible] I tell you what the Boys are all down on General Hunter any way he is to

[3] dropped anchor and pepered away and they piped all hands down to dinner they all went down and let the Fort fire away she got struck a good many times but it did not hert her any only made a few dents in her we all expect to have a darn hard fight when we do git at Charleston the Pickits were attack over on the other side that is Hilton Head last night and one of the Cavalry was killed I believe

I got your last papers of the 4 of March last Wednesday they was the only ones for the Company there was quite a lot of news they had to go the rounds through the Company I see they got the news in about General Steavenson and the Mass troops in it I tell you what the Boys are all down on General Hunter [?] any way he is the

Adelbert A. Buffum.

Port Royal S.C. Mch 15 1863 Aug. 28

Homliest man out of jail
I believe [illegible] some expect
Old Brown's back to take
command [illegible] expedition
when General [illegible] was
[illegible]
camp and [illegible] he got 3 belly
cheers [illegible] when
he went [illegible] he got 3 more
I don't get many letters lately [illegible]
how [illegible] this
I wish you would direct it to
Port Royal I believe you did
direct [illegible] papers
and [illegible]
I got them [illegible]
You must excuse this writing for
I have got no good place to write
with ink I have to sit on the
ground and [illegible] knees for a writing
[illegible] my love to all from Dell

[4] Homliest man out of jail I believe us fellers some expect Old Burnie back to take command of this Expedition when General Steavenson was [pencil is very blurred and faded because it's on a crease; perhaps "resigned he came right up to our"] camp and did et he get 3 bully cheeres well I gusse he did and when he went over to the [10?] Co he got 3 more

I dont get many letters latly some how or other when you answer this I wish you would direct it to Port Royal I believe you did direct it so on the last papers and I suppose that is the reson that I got them so quick

you must excuse this writeing for I have got no good place to write with ink so I have to sit on the ground and use my knees for a writeing desk

my love to all from Dell

[Endorsed in CAB's handwriting Adelbert A. Buffum

Port Royal S.C.

Mch 15 1863

Ans. 28th]

Seabrook Island, S.C., April 22, 1863

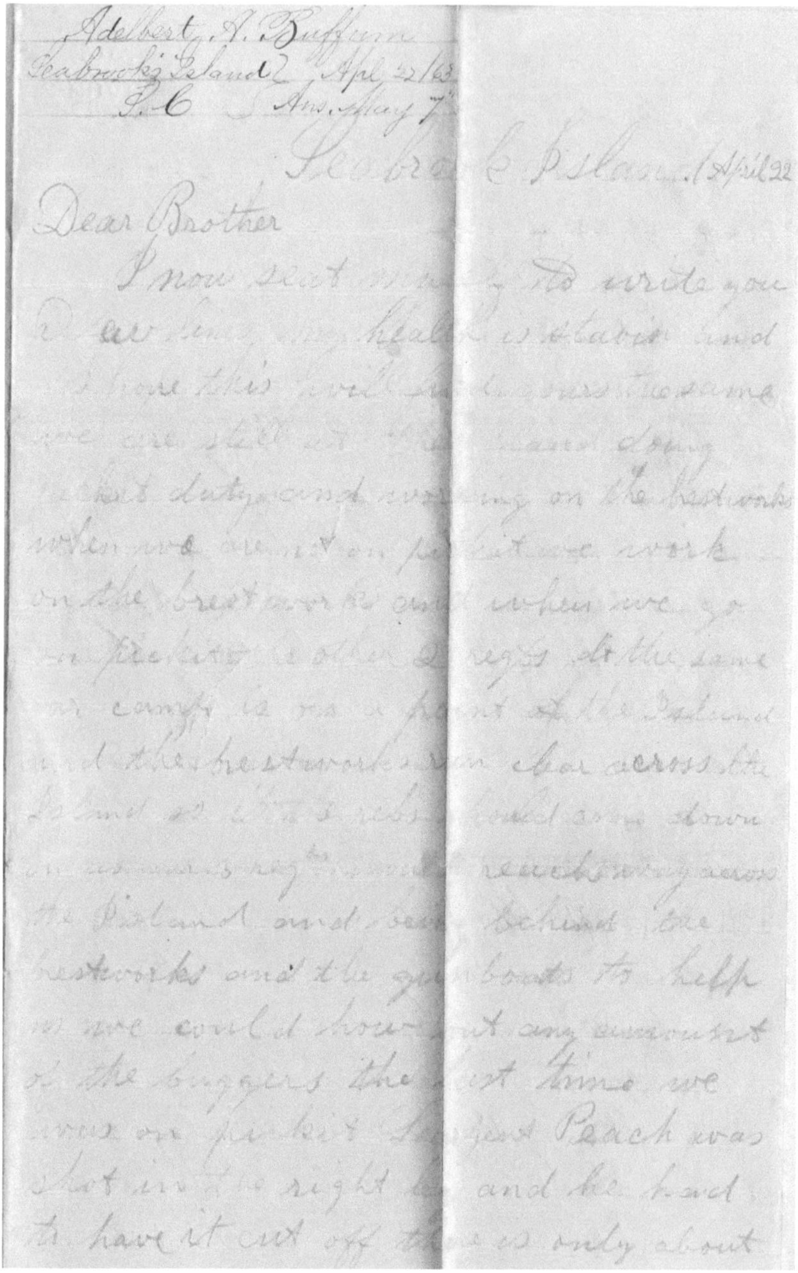

Adelbert A. Buffum
Seabrook Island Apl 22/63
S.C. Ans. May 7

Seabrook Island April 22

Dear Brother

I now seat myself to write you a few lines my health is about [illegible] and I hope this will find you the same we are still at this island doing picket duty and working on the breastworks when we aren't on picket we work on the breastworks and when we go on picket the other 2 regts do the same our camp is on a point of the island and the breastworks run clear across the island so that if the rebs should come down [illegible] reg[illegible] reaching way across the island and being behind the breastworks and the gunboats to help us we could hold out any amount of the buggers the last time we were on picket Lieutenant Peach was shot in the right leg and he had to have it cut off there is only about

[In pencil and faded

Endorsed in CEB's hand

"Adelbert A. Buffum

Seabrook's Island S.C.

Apl 22/63

Ans. May 7"]

Seabrook Island April 22 [1863]

Dear Brother

I now seat myself to write you a few lines, my health is stavin and I hope this will find yours the same we are still at the island doing pickit duty and working on the brestworks when we are not on pickit we work on the brestworks and when we go on pickit the other regts do the same our camp is on a point of the Island and the brestworks run clear across the Island so it the regs should come down on us our 3 regts would reach way across the Island and being behind the brestworks and the gunboats to help us we could how out any amount of the buggers the last time we was on pickit Seargent Peach was shot in the right leg and he had to have it cut off there is only about

4 inches of the stub left the next day after we was relieved we was pay off and we got up a subscription paper for Peach and most evry man in the company gave something he got about 800 dollars from our company and some of the officers gave something and there was some sinet from evry company in the regt so in all he got about 600 dollars and as he was a poor man it was quite a help to him he is on board of the hospital ship doing well
The morning that we was relieved Co B & C was called up at 2 oclock the morning and we came down to camp and stacked arms on the beach and took our things of and struck all the tents in the regt and carried them down to the landing we got every thing down before day light so when the Regt came down they could go right on board of the boat & cam

[2] 4 inches of the stub left the next day after we was relieved we was pay off and we got up a subscription paper for Peach and most evry man in the company gave something he got about 300 dollars from our company and some of the officers gave something and there was some sined from evry company in the regt, so in all he got about 600 dollars and as he was a poor man it was quite a help to him he is on board of the hospital ship doing well

The morning that we was relieved Co B & C was called up at 2 oclock in the morning and we came down to camp and stacked arms on the beach and took our things of and struk all the tents in the regt and carried them down to the landing we got every thing down before daylight so when the Regt came down they could go right on board of the boat because we had orders to start for Newbern and I tell you we was all glad of it for Newbern seemed like a home to us we have not seen it in 6 months we heard that Generel Foster was in a tight place and so we were going to reinforce him and as the monitors had all drawn away from Charleston and old Hunter had recalled his troops we was glad to get out of the place well the rest of the Regt got down at 10 oclock and we layed round all day untill 5 oclock when the order was countermanded and we had orders to bring the tents back again I tell you there was some tall swearing in the Regt. about that time but there was no help for it and we had to bring all that we stuff back that we had luged down and pitch camp again and most of the boys had burned there bedding so we had to go and cut more but

[3] we had orders to start for Newbern and I tell you we was all glad of it for Newbern seemed like a home to us we have not seen it for 6 months we heard that General Foster was in a tight place and so we were going to reinforce him and as the monitors had all drawn away from Charleston and old Hunter had recalled his troops we was glad to get out of the place well the rest of the Regt got down at 10 oclock and we layed round all day untill 5 oclock when the order was countermanded and we had orders to bring the tents back again I tell you there was some tall swearing in the Regt about that time but there was no help for it and we had to bring all that we stuff back that we had lugede down and pitch camp again and most of the boys had burned their bedding so we had to go and cut more but

that did not take: us long as we
sleep on bulrushes and there is plenty of
them grow here now they have but a stop
to our geting any more oysters as we have
a roll call evry 2 hours of the day and are
not allowed to go out of the camp
The talk is they are going to have
another slap at Charleston there is 4 of
the monitors here again and we expect
11 more of them 7 new ones they mean
to go at Charleston in ernest this time
by thunder I hope they will do better
than they did before I will send
you some Palmetto seeds that I got off
from a Palmetto tree but I dont suppose
they would grow there if you should plant
them but it would do no hurt to try well
I must close as I have not got
any more time now my love to all
from your Brother Del.

P S Come to look for thoes seeds I find
I have lost all except 2 there is no
Palm trees in camp I shall have to
wait till I go on picket again when
I will send you some more

from D. Buffum.

[4] that did not take us long as we sleep on bullrushes and there is plenty of them grow here now they they have put a stop to our geting any more oysters as we have a roll call every 2 hours of the day and are not allowed to go out of the camp

The talk is they are going to have another slap at Charleston there is 4 of the monitors here again and we expect 11 more of them 7 new ones they mean to go at Charleston in ernest this time by thunder I hope they will do better than they did before I will send you some Palmetto seeds that I got off from a Palmetto tree but I dont suppose they will go there if you should plant them but it would do no hurt to try well I must close here for I have not got any more time now my love to all from your Brother Dell

PS Come to look for thoes seeds I find I have lost all except 2 there is no Palm trees in camp I shall have to wait till I go on pickit again when I will send you some more

from A Buffum

Camp Dupont, Seabrook Island, S.C., June 16, 1863

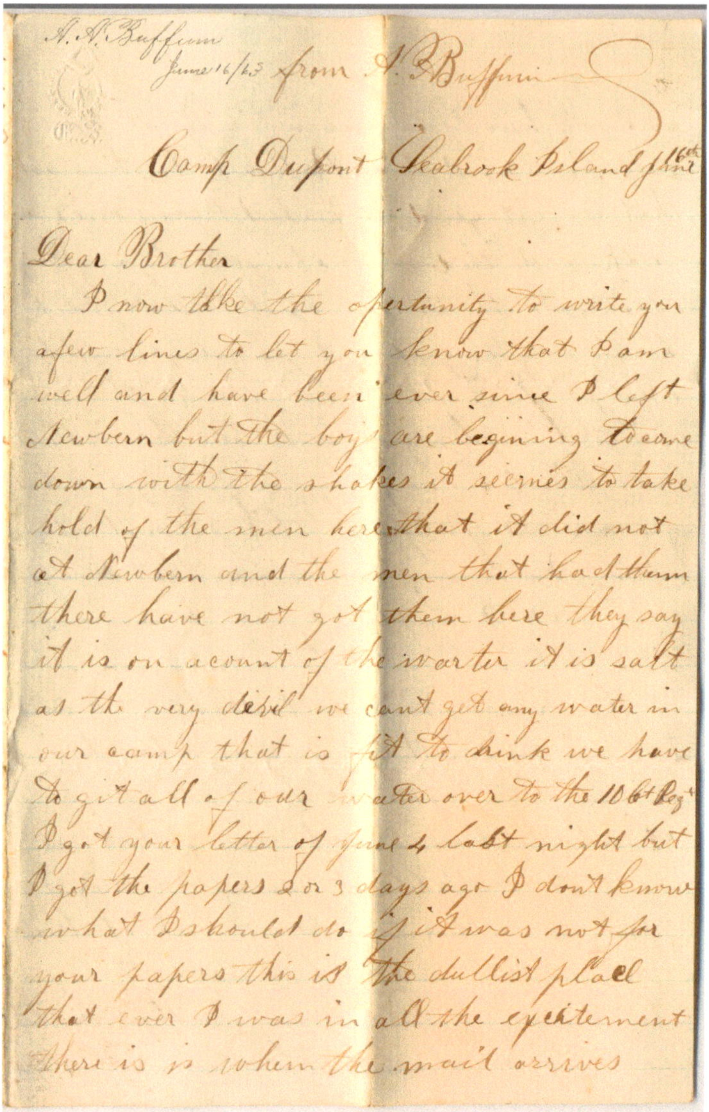

A. A. Buffum
June 16/63 from A. A. Buffum

Camp Dupont Seabrook Island June 16th

Dear Brother
I now take the opertunity to write you
a few lines to let you know that I am
well and have been ever since I left
Newbern but the boys are begining to come
down with the shakes it seemes to take
hold of the men here that it did not
at Newbern and the men that had them
there have not got them here they say
it is on acount of the warter it is salt
as the very devil we cant get any water in
our camp that is fit to drink we have
to git all of our water over to the 106t Regt
I got your letter of June 4 last night but
I got the papers 2 or 3 days ago I dont know
what I should do if it was not for
your papers this is the dullist place
that ever I was in all the excitement
there is is when the mail arrives

Endorsed in CEB's hand A.A. Buffum June 16/63. Is it 1863?]

Camp Dupont Seabrook Island June 16th.

Dear Brother

I now take the opertunity to write you a few lines to let you know that I am well and have been ever since I left Newbern but the boys are beginning to come down with the shakes it seemes to take hold of the men here that it did not at Newbern and the men that had them there have not got them here they say it is on acount of the water it is salt as the very devil we cant get any water in our camp that is fit to drink we have to git all of our water over to the 106t Regt I got your letter of June 4 last night but I got the papers 2 or 3 days ago I dont know what I should do if it was not for your papers this is the dullist place that ever I was in all the excitement there is is when the mail arrives

and we have bloody mean libing we have
salt horse and boiled Pork all the time all
that we git good is Baked Beans once a
weedk and there is no chance to buy anything
except at the sutlers and he dont keep
hardly anything except paydays and he
charges just what he wants to
but there is plenty of work here there is 150
men detailed every day for fatigue I tell
you we are fortifying this Island putty
strong. I am very sorry to hear that Alonzo
is so sick I suppose it will be a long
time before he will be round again
so Oscar wants to go to Boston to work
I think he had better stay where he is
or else go to war we need all shuch
stough fellows in the army that we
can get so Mr Lord has left Father I
thought Father would get sick of him
he is another man that ought to be in
the army you did not say wether Father
had eny help or not I suppose Zede is

[2] and we have bloody mean living we have salt horse and boiled [?] Pork all the time all that we git good is Baked Beans once a weeak and there is no chance to buy anything except at the sutlers and he dont keep hardly anything except paydays and he charges just what he wants to but there is plenty of work here there is 150 men detailed every day for fatuigue I tell you we are fortifying this Island pretty strong. I am very sorry to hear that Alonzo is so sick I suppose it will be a long time before he will be sound again so Oscar wants to got to Boston to work I think he had better stay where he is else go to war we need all shuch stout fellows in the army as we can get so Mr Lord has left Father I thought Father would get sick of him he is another man that ought to be in the army you did not say wether Father had any help or not I suppose Tede is

big enough to go to work but he ought to go to school some time yet I suppose but I dont see how father can get along without help tell him I will be home to help in a year if the story is true about all the first quoto of three years troops are to be discharged the first of next June but if the war should not be over then I suppose that I should enlist again I should like to see all such fellows as Fountain and Seitch drafted there is no fellow in this regt that I used to know but Bill Wyman and he is well you said something about your being 21 next year and then you should have to leave your place what do you mean did your boss only engage you for so long a time or how is it so you will be 21 next April then I suppose I will be 18 next August to tell the truth I have almost forgot how old I was as I have been passing for 19 or 20 all the time I set out to write and ask Tedes age and then I would

[3] big enough to go to work but he ought to go to school some time yet I suppose but I dont see how father can get along without help tell him I will be home to help in a year if the story is true about all the first quota of three years troops are to be discharged the first of next June but if the war should not be over then I suppose that I should enlist again I should like to see all such fellows as Fountain and Leitch drafted there is no fellow in this regt that I used to know but Bill Wyman and he is well you said something about your being 21 next year and then you should have to leave your place what do you mean did your bos only engage you for so long atime or how is it so you will be 21 next April then I suppose I will be 18 next August to tell the truth I have almost forgot how old I was as I have been passing for 19 or 20 all the time I set out to write and ask Tedes age and then I would

be shure of my age as he is just 2 years
younger than I am I had an Idea in my
head that I was 19 in August I want
you to send me your picture as I have got all
the family except you and Leony and Lester
I want you to get yours and Addies
and send them along and send Tommys
when he gets well enough to have it taken
Sergeant Cobb and Corporal Norris have
had there furloughs in there pockets for
over a month and they got started and got
as far as Hilton Head once and were sent
back and have been here ever since
I suppose I could get a furlough if I had
some rich and influenchal friends at home
as some of the men have there is any
quantity of Sergeants and Corporals gone
home to take commishions in the nigger
Regts at home there is some gone from
evry company but this one Cobb and
Willey both say they can get a chance if
they can get home but I dont want to
loose them they are such bloody good fellows
there has been a number of promotions in this
regt latly we have got a little snot of a Lieut
in our company that used to be Orderly of C.
he has been promoted over our Orderly they dont
do things fair in this Regt no way as I have
got down to the bottom of the page I must dry up
from your Brother dell

tell Father if he does not send the box that I wrote for
before you get this that he can send a pair of slippers
his a ...

[4] be shure of my age as he is just 2 years younger than I am I had an Ide[a] in my head that I was 19 in August I want you to send me your picture as I have got all the family except you and Lony and Sis so I want you to get yours and Addies and send them alone and send Lonys when he gets well enough to have it taken. Sergeant Cobb and Corporal Norris have had there furloughs in there pockets for over a month and they got started and got as far as Hilton Head once and were ordered back and have been here ever since I suppose I could get a furlough if I had some rich and influnetial friends at home as some of the men have there is any crist[?] quanaty of Sergeants and Corporals gone home to take commishions in the nigger Regts at home there is some gone from evry company but this one ~~it [unreadable]~~ Cobb and Willey both say they can get a chance if they can get home but I dont want to loose them they are such bloody good fellows there has been a number of promotions in this regt latly we have got a little smot [?] of a Leiut in our company that used to be Orderly of Co C [?] he has been promoted over our Orderly they dont do things fair in this regt no way as I have got down to the bottom of the page I must dry up

from your Brother dell

[crossed]Tell Father that if he does not send the box that I wrote for before you get this that he can send me a pair of slippers if he has a mind to

MORRIS ISLAND, S.C., SEPTEMBER 5, 1863

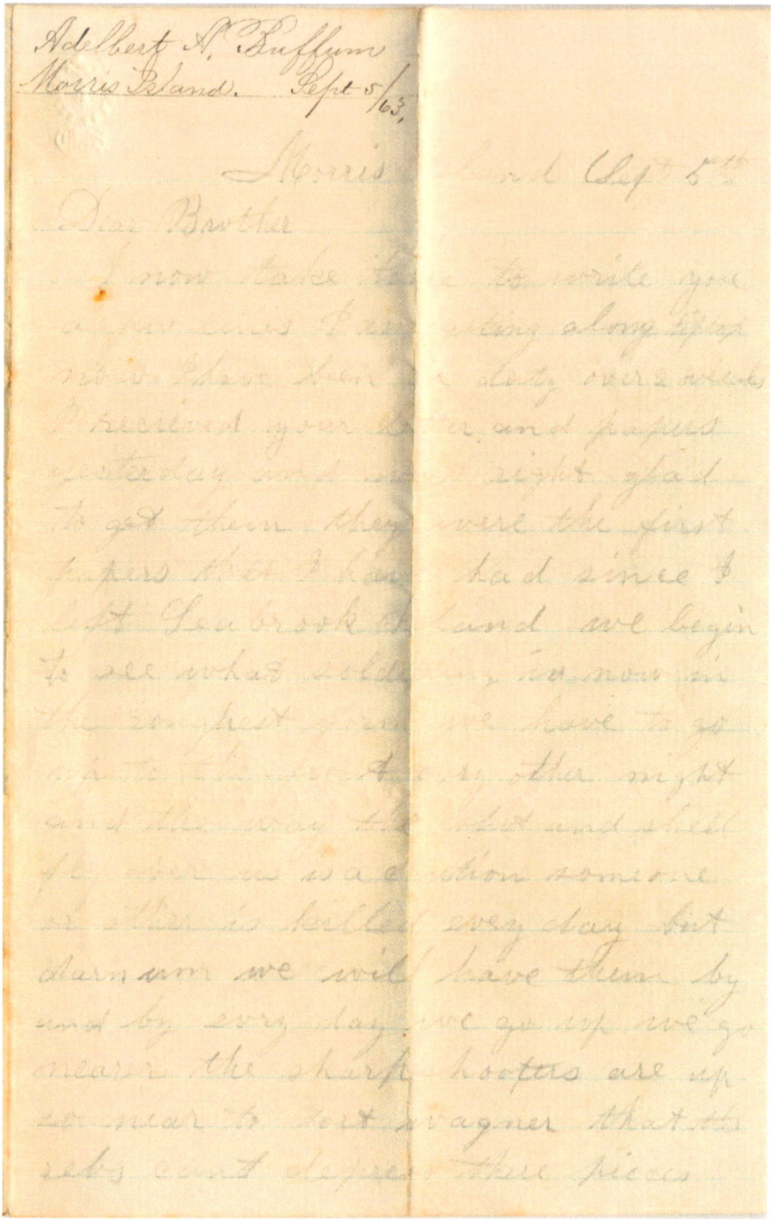

Adelbert A. Buffum
Morris Island. Sept 5/63,

Morris [illegible] Sept 5th
Dear Brother
I now take [illegible] to write you
a few lines I am [illegible] along tip top
[illegible] have been [illegible] duty over 2 week
I received your [illegible] and papers
yesterday and [illegible] right glad
to get them they were the first
papers that I have had since I
left Seabrook Island we begin
to see what soldiering is now in
the roughest [illegible] we have to go
up to the front every other night
and the way the shot and shell
fly over us is a caution someone
or other is killed every day but
darn um we will have them by
and by every day we go up we go
nearer the sharp shooters are up
so near to Fort Wagner that the
rebs cant depress their pieces

[Endorsed in CEB's hand

"Adelbert A. Buffum

Morris Island. Sept 5/63."]

Morris Island Sept 5th.

Dear Brother

I now take time to write you a few lines I am geting along tiptop now I have been on duty over 2 weeks I recieved your letter and papers yesterday and was right glad to get them they were the first papers that I have had since I left Seabrook island we begin to see what soldiering is now in the roughest way we have to go up to the front every other night and the way the shot and shell fly over us is a caution someone or other is killed eve[r]y day but darn um we will have them by and by evry day we go up we go nearer the sharp shooters are up so near to Fort wagner that the rebs cant depress there pieces

enough to do them any harm and
the front of old Sumter looks
like a pile of old brick and morter
the marcins go almost up to Sumter
evry night in small boats on
picket I tell you that Charleston
is sure to fall this time. this
siege work is slow but shure
we can stand up to our sink and
see the 200 pounders plug sumter and
most evry shot makes the brick
fly Sumter has not returned a shot
for a long time I tell you we have
got quite used to the fireing we hear
it every hour of the day we have
got so used to it that we can dodg
most evry shot in the day time we
lay behind our brest works and
our fire and the rebel fire go over
us I used to lay and watch the shot
go over me I would not believe it till
I saw them but we can see evry

[2] enough to do them any harm and the front of old Sumter looks like a pile of old bricks and mortar the mareine[s] go almost up to Sumter evry night in small boats on pickit I tell you that Charleston is sure to fall this time. This seige works is slow but shure we can stand up to our sink and see the 200 pounders plug sumter and most evry shot makes the brick fly Sumter has not returned a shot for a long time I tell you we have got quite used to the fireing we here it every hour of the day we have got so used to it that we can dodg most any shot in the day time we lay behind our brest works and our fire and the rebel fire go over us I used to lay and watch the shot go over us I would not believe it till I saw them but we can see evry

shot we we are where we can
see them leave the gun when the
shot is in the air it looks like a
foot ball and you can see them as plain
as day and in the night time the shot
look like a shooting star going
through the sky our regt made a
charge the other night and they say
it was as good a one as ever was made
we charged on the rebel rifle pits and took
them you ought to see the rebs skedadle
mud up to there knees and they going
it like thunder but our boys dropped a
good many of them before they reached
the [illegible] we captured 84 of them since
then we have got [illegible] works way
out beond as far as they can go on
acount of the marsh tell Littlefield
that Leint Perkins of Co. I was killed
in that charge and Leint Trout
of Co. D has been promoted to first Leint
in his place and Sargent Foster of

[3] shot we [duplicated "we"] are where we can see them leave the gun when the shot is in the air it looks like a foot ball and you can see them as plane as day and in the night time the shot look like a shooting star going through the sky our regt made a charge the other night and they say it was as good a one as ever was made we charged on the rebel riflepits and took them you ought to see the rebs skedadle mud up to there knees and they going it like thunder but our boys dropped a good many of them before they reached Wagner we captured 84 of them since then we have got brestworks way out beond as far as they can go on acount of the marsh tell Littlefield that Leiut Perkins of Co. I was killed in that charge and Leiut Sweet of Co D has been promoted to first Leiut in his place and Sargent Foster of

Co H has been promoted to 2nd Lieut
and we expect our Orderly is to be 2nd
Lieut of Co D and you no need to look
on the soldiers caps to see if you [illegible]
24 on them for there is no furlough to be
given out of this regt yet a while that
order was countermanded before we
left Falmouth [illegible] [illegible] [illegible] [illegible]
went home and they went with
Lieut Col Stevenson for concripts [illegible]
to fill up our regt there is a good
many sick in our regt now there
is more go to [illegible] call now than
ever before since we left home
we go on picket to night we go
every other night and come in at day [illegible]
this change [illegible] [illegible] [illegible] [illegible] [illegible]
[illegible] [illegible] [illegible] [illegible] [illegible]
that change now we only have to stay 12
hours and before we had to go at dark and
stay 24 hours day and night so we get
rid of laying [illegible] in the hot sun [illegible]
day I dont know as I can write
any more now but will try and write
oftener than I have been writing
give my love to [illegible] [illegible] and [illegible]
the folks and write soon from D C

[4] Co H has been promoted to 2nd Leiut and we expect our Orderly is to be 2nd Leuit of Co D and you no[w] need to look on the soldiers caps to see if you can [find] 24 on them for there is no furloughs to be given out of this regt yet a while that order was countermanded before we left Sebrook there was [unreadable; "only 1 or 2"?] went home and they went with Leuit Col Stevenson for concripts, for to fill up our regt there is a good many sick in our regt now there is more go to surgoen call now than ever before since we left home we go on pickit to night we go on evry other night and come in at daylight this chance was given us as an honor by General Gillmore for the good we done on that charge now we only have to stay 12 hours and before we had to go at dark and stay 24 hours day and night so we git rid of laying out in the hot sun all day I dont know as I can write any more now but will try and write oftener than I have been writeing give my love to Littlefield and all the folks and write soon from Dell

MORRIS ISLAND, S.C., SEPTEMBER 12, 1863

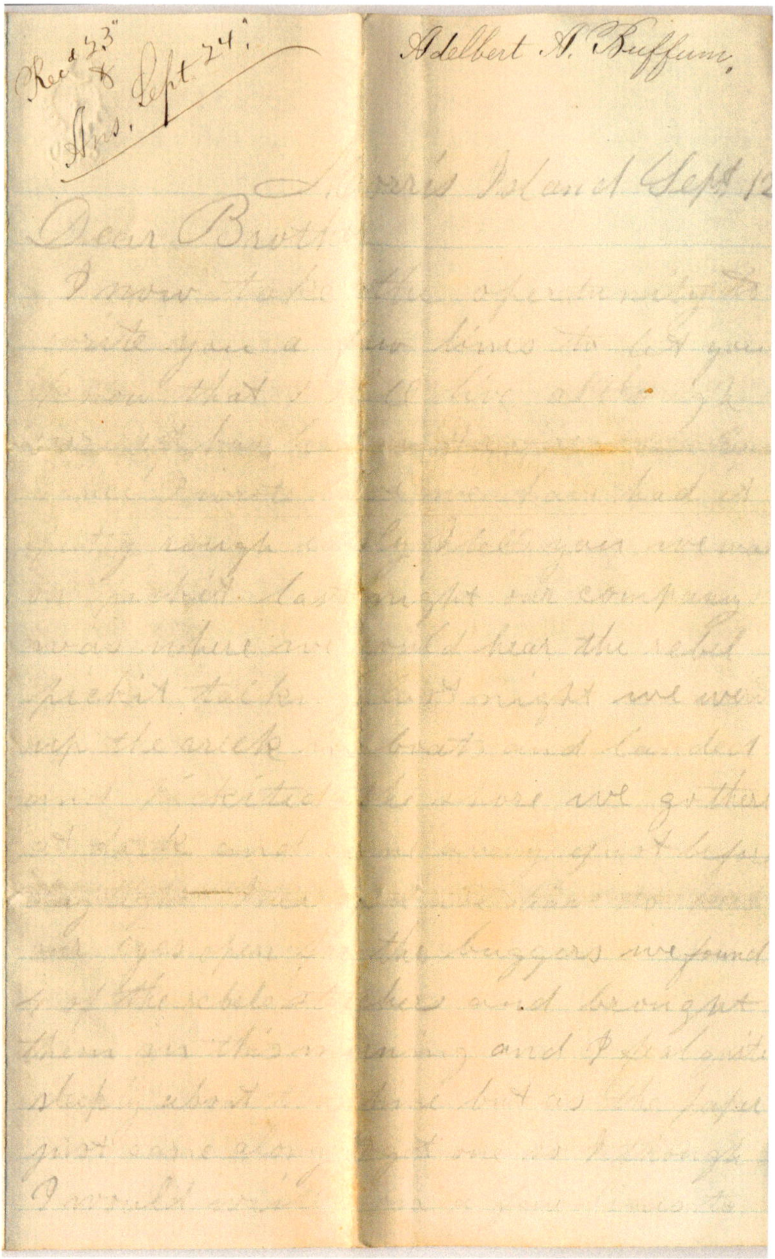

Rec'd 23" 8

Ans. Sept 24.

Adelbert A. Buffum,

Morris Island Sept 12

Dear Brother

I now take the opportunity to write you a few lines to let you know that I [illegible] alive altho [illegible] [illegible] [illegible] [illegible] [illegible] had a pretty rough [illegible] I tell you we was on picket last night our company was where we could hear the rebel picket talk [illegible] night we went up the creek [illegible] boats and landed and picketed the shore we got there at dark and [illegible] away just before daylight [illegible] [illegible] eyes open [illegible] the buggers we found [illegible] of the rebels [illegible] and brought them in this morning and I [illegible] sleep about [illegible] but as the paper just came along [illegible] I thought I would [illegible] a few lines to

[Endorsed in CEB's hand

"Rec'd 23 & Ans. Sept 24.

Adelbert A. Buffum"]

Morris Island Sept. 12

Dear Brother

I now take the opertunity to write you a few lines to let you know that I still live although our regt has been under fire [unreadable; "nearly"?] since I wrote last we have had it pretty rough lately I tell you we was on pickit last night our company was where we could hear the rebel pickit talking last night we went up the creek in boats and landed and pickited the shore we go there at dark and come away just before daylight I tell you we have to keep our eyes open for the buggers we found 4 of the rebels strechers and brought them in this morning and I feel quite sleeply about this time but as the paper just came along I got one so I though[t] I would write on a few lines to

send with it I dont feel like
writing a very long letter for I don
know but we will be ordered out to
night for some duty or other but there
one thing shure that we have not
got Fort Wagner to charge now that
what the boys have been dreading all
along but now she's ours as is the
whole Island and so will the city
of Charleston be before 2 weeks I
believe the rebs get the devil all
the time on Sullivans Island now
from the fleet, in this paper I send
you it tells you about the charge
our Regt made a while ago on the
rebels rifle pits which we were on the
charge before we got to their works there
was a man shot not more than 2 yards
from me and as I passed him on
the run I looked down and saw the
blood on his head and I knew he
was dead he was John Burns

[2] send with it I dont feel like writeing a very long letter for I dont know but we will be ordered out to night for some duty or other but theres one thing sure that we have not got Fort Wagner to charge now thats what the boys have been dreading all along but now shes ours as is the whole Island and so will the city of Charleston be before 2 weeakes I believe the rebs catch the divel all the time on Suleivans Island now from the fleet in this paper I send you it tells you about the charge our Regt made a while ago on the rebels rifle pits when we were on the charge before we got to there works there was a man shot not more than 3 yards from me and as I pased him on the run I looked down and saw the blood on his head and I knew he was dead he was John Burns

of Co. E instead of Co. D as the
paper states all the rest was
killed and wounded after we got
to the rebes works that was the first
regular charge we ever was in and
I hope it will be the last for these
charges are pretty rough things to be
in I tell you. Alva Clontman is
in our Co. he came since we have
been on this island he used to work
for Martin in the boss I suppose you
know him Teeder does I know he was
in the charge and the way them long
legs of his took him over the ground
was a caution he makes a bully
soldier I tell you our orderly expects
to go home on a furlough next wednesday
and he is going to call round at the
store he has got an uncle or some other
relation at work at Bill Alens shoe
store in dock square Orderly expects
a commishion in the 54 regt but the

[

3] of Co E instead of Co I as the paper states all the rest was killed and wounded after we got to the rebes works that was the first regular charge we ever was in and I hope it will be the last for these charges are pretty rough things to be in I tell you. Alva Cloutman is in our Co, he came since we have been on this island he used to work for Martin in Melrose I suppose you know him Tede does I know he was in the charge and the way them long legs of his took him over the ground was a caution he makes a bully soldier I tell you our orderly expects to go home on a furlough next wednesday and he is going to call round at the store he has got an uncle or some other relative at work at Bill Alens shoe store in dock square Orderly expects a comishion in the 54 regt but the

Col toled him that he was the next chance for promotion in the regt if he stays he will be 2nd Lieut of Co D. when we was coming back from Wagner the other morning when we was expecting to leave to charge that place I saw Littlefield that used to live at Mr Tucker's in Melrose he is 2nd Lieut in the 54 Regt I did not have a chance to talk with him but we saw each other and bowed as we were marching along the last time I saw him he was a private in the 44 Regt at Newbern I recieved a letter from Mother to day and six papers I suppose that was from you as it was your hand writeing I shall answer Mothers letter as soon as I get a chance I must close here you must excuse this poor short letter for I am darnnation sleepy for I was up all night last night and I had to clean my gun this forenoon and we dont know when we shall be ordered out so we get sleep when we can give my love to all the folks from your affectionate Brother Dell Buffum

[4] Col toled him his was the next chance for promotion in this regt if he stays he will be 2nd Leuit of Co D. When we was comeing back from wagner the other morning when we was expecting to have to charge that place I saw Littlefield that used to live at Mr Tucker's in melrose he is 2nd Leiut in the 54 Regt I did not have a chance to talk with him but we saw each other and bowed as we were marching along the last time I saw him he was a private in the 44 Regt at Newberg I recieved a letter from Mother to day and six papers I suppose that was from you as it was your hand writeing I shall answer Mothers letter as soon as I get a chance I must close here you must excuse this poor short letter for I am darnnation sleepy for I was up all night last night and I had to cleal [clean?] my gun this forenoon and we dont know when we shall be ordered out so we get sleep when we can give my love to all the folks from your affectionate Brother Dell Buffum

St. Augustine, Florida, Nov 2nd 1863

St Augustine Fla Nov 2nd 1863

Dear Brother

I think it is about time I wrote
to you but I have been sick lately and
[illegible] dont feel like [illegible] but I am
getting better now and will write oftener
I recieved a letter and some papers the day
before yesterday from you and in it
you thought I would be home soon
but I think it will be some time before
I get there I shall be [illegible] by next September
if I live the 10th Ct Regt has come down
here and it makes our duty considerable
easyer we was payed off yesterday and
now we intend to [illegible]

so you have been up to New Hampshire
I should have thought they could have
afforded to give you a longer vacation
after working for so long steady
I suppose Ede is most large enough to
go to work at the store if it were not

St. Augustine Fla Nov 2nd 1863

Dear Brother

I think it is about time I wrote to you but I have been sick lately and did not feel like writeing but I am geting better now and will write oftener I recieved a letter and some papers the day before yesterday from you and in it you thought I would be home soon but I think it will be some time before I get there I shall be there by next September if I live the 10 Ct Regt has come down here and it makes our duty conciderable easyer we were payed off yesterday and now we intend to live

So you have been up to New Hampshier I should have thought they could have afforded to give you a longer vacation after working for so long steady I suppose Tede is most large enough to go to work at the store if it were not

for his schooling I suppose Luther wou[ld]
have him long ago it is best for him
to go to school as long as he can
Sergeant Cobb is our Orderly now as
Chipman is discharged from the company
we have eny quantity of good fruit here
called the Guava it is stavin fruit I tell
you it is what made me get better of the
diarahea so quick I would give 5 dollars
if I could get some guavas home shuch as
I had today but I cant it rots so quick
we have plenty of oranges here the sour
ones are so plenty that the boys pick
them off and throw them at each other
round the street there is lots of fish here
the trout and bass and mullet are
very plenty.
Nov 4th Since writeing the above I have
been on provo guard and as we were
patroleing past the picture gallery I says
to the fellow that was with me lets go
in and have our pictures taken so we went
and got them taken with guns and

[2] for his schooling I suppose father would have him long ago it is best for him to go to school as long as he can

Sergeant Cobb is our Orderly now as Chipman is discharged from the company we have eny quanteity of good fruit here called the Guava it is staving fruit I tell you it is what made me get better of the diarahea so quick I would give 5 dollars if I could get some guavas home shuch as I had today but I cant it rots so quick we have plenty of oranges here the sour ones are so plenty that the boys pick them off and throw them at each other round the street there is lots of fish here the trout and bask and bullit are very plenty.

Nov 4th Since writeing the above I have been on provo guard and as we were patroleing past the picture gallery I sayes to the fellow that was with me lets go in and have our pictures taken so we went and got them taken with guns and equipments I will send you mine and I want you to get yours taken and send me you can see how poor I am in this picture I am 32 lbs lighter than I was on Seabrook when I was there I weighed 147 pounds and now I weigh only 115 but I am gaining now and I guess I will thicken up again I think I will go down and have another taken without equipments sometime when I am flush I will send you a New South but I dont suppose there is any news to you in it but as long as I have got it I will send it, you can tell Mother that I am on full duty again now

We have a theatre here the 48th regt had some seanery and fixings and our officers bought it of them for 300. dollars and now they have a performance twice a weak we have got some very good actors in our regt and they get up very good plays we have to get a pass from the Capt singed by the Col to go the front

[3] equipments I will send you mine and I want you to get yours taken and send me you can see how poor I am in this picture I am 32 lbs lighter than I was on Seabrook when I was there I weighed 147 pounds and now I weigh only 115 but I am gaining now and I gusse I will thicken up again I think I will go down and have another taken without equipments sometime when I am frush I will send you a New Youth [South?] but I don't suppose there is any news to you in it but as long as I have got it I will send it, you can tell Mother that I am on full duty again now

We have a theatre here the 48th regt had some seanery and fixings and our officers bought it of them for 300. dollars and now they have a performance twice a weak we have got some very good actors in our regt and they get up very good plays we have to get a pass from the Capt singed by the Col to go the front seats are for the officers and ladies and the privates have to take the back seats I have been twice and the house was crowded both times the house will hold 500 people the drummer boys dress up and take the part of weomen and they take there part tip top & well I have not got time to write any more now as I have just come of guard and have got my things to clean up so I will close with my love to all

from your Affectionate Brother Dell

P.S. Dont forget to send me your picture in the next letter

from Dell

Adelbert,
St. Augustine Fla.
Nov. 2, 1863

[4] seats are for the officers and ladies and the privates have to take the back seats I have been twice and the house was crowded both times the house will hold 300 people the drumer boys dress up and take the part of women and they take there part tip top to well I have not got time to write any more now as I have just come off guard and have got my things to clean up so I will close with my love to all from your Affectionate Brother Dell

P.S. Dont forget to send me your picture in the next letter

from Dell

St Augustine, Florida, November 28, 1863

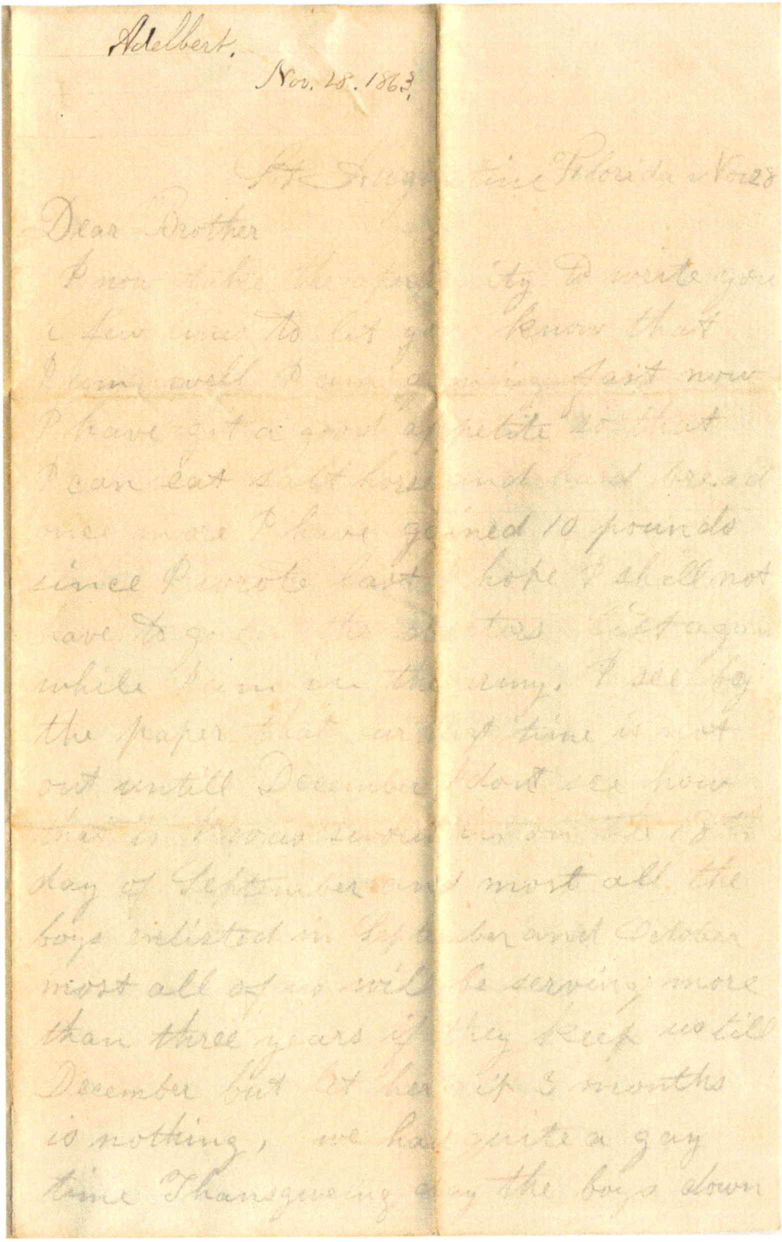
Adelbert.
Nov. 28. 1863.

St Augustine Florida Nov 28
Dear Brother
I now take the opportunity to write you
a few lines to let you know that
I am well I am gaining fast now
I have got a good appetite so that
I can eat salt horse and hard bread
once more I have gained 10 pounds
since I wrote last I hope I shall not
have to go in the [illegible] [illegible] again
while I am in the army. I see by
the paper that our [illegible] time is not
out untill December I dont see how
that is I was sworn in on the 18th
day of September and most all the
boys enlisted in September and October
most all of us will be serving more
than three years if they keep us till
December but let here it 3 months
is nothing, we had quite a gay
time Thanksgiving day the boys down

[Endorsed "Nov. 28. 1863,"]

St. Augustine Florida Nov. 28

Dear Brother

I now take the opertunity to write you a few lines to let you know that I am well I am gaining fast now I have got a good appetite so that I can eat salt horse and hard bread once more I have gained 10 pounds since I wrote last I hope I shall not have to go on the doctors list again while I am in the army. I see by the paper that our Regt time is not out untill December I dont see how that is I was sworn in on the 18th day of September and most all the boys enlisted in September and October most all of us will be serving more than three years if they keep us till December but let her rip 3 months is nothing, we had quite a gay time Thansgiving day the boys down to the fort got up a greased Pig whoever caught him had him and a greased pole with a 5 dollar on the top for eny one who could clime to it and a sack race the prize being 5 dollars and then there was a boat race and then came the target shooting the 3 best shots picked out of the companies at the fort with a prize of 5 dollars to the man that made the best shot both with the muskit and the cannon and after that was over we went out in the field and and kicked foot ball and played base ball for dinner we had beafsteak and sweet potatoes and plum pudding made out of soaked hard bread with a few raisins and spices in it and for supper we had nuts and raisins and Oranges after supper some of the boys blacked up with burnt cork and there was one that played the boans and another played the fiddle

[2] to the fort got up a greased Pig whoever caught him had him and a greased pole with a 5 dollar on the top for eny one who could clime to it and a sack race the prize being 5 dollars and then there was a boat race and then came the target shooting the 3 best shots picked out of the companies at the fort with a prize of 5 dollars to the man that made the best shot both with the muskit and the cannon and after that was over we went out in the field and and kicked foot ball and played base ball for dinner we had beafsteak and sweet potatoes and plum pudding made out of soaked hard bread with a few raisins and spices in it and for supper we had nuts and raisins and Oranges after supper some of the boys blacked up with burnt cork and there was one that played the boans and another played the fiddle

and the rest of us danced on the whole we had a very good Thanksgiving I have not had a letter from home for the last 2 mailes it is about time I had a letter with yours and Tesles pictures I dont know as I can think of enything more to write about when I tell you that I am well I tell you all there is to tell all there is going on here is guard duty and inspection and dress parade Thansgiving night I went to a concert given by the 24 Glee Club they are singers picked out of the regt and they make the best singing that ever I heard the Church that they had the concert in was built in 1775 I wish you was here to look round and see all the old buildings and the old fort there is some old guns in the fort that the Spanish brought over when they settled the place and there is dark dungeons and out from one of these is a small round

[3] and the rest of us danced on the whole we had a very good Thansgiveing

I have not had a letter from home for the last 2 mails it is about time I had a letter with yours and Tedes pictures I dont know as I can think of enything more to write about when I tell you that I am well I tell you all there is to tell all there is going on here is guard duty and inspection and dress parade Thansgiveing night I went to a concert given by the 24 Glee Club they are singers picked out of the regt and they make the best singing that ever I heard the Church that they had the concert in was built in 1775 I wish you was here to look round and see all the old buildings and the old fort there is some old guns in the fort that the Spanish brought over when they settled the place and there is dark dungeons and out from one of these is a small round

hole where an indain chief called the
Wild Cat cut his way out through the
stone like fort with a knife this Wild
Cat was a Chief that they had prisoner
in the Florida war he escaped from
them 3 times and the last time he got
away he was most starved and naked
and they found his dead body in the
woods a few days after there is an old
gray headed man that lives near the
barracks that tells us meny a rough
old tale about that old fort it has
been scaled twice by the indains and
they were driven back bodth times with
an awfull slaughter I tell you there
has been some rough fighting in
this place there is the history of this
place in the post office but it can't be
got for love nor money as it is the only
one round here it is near as large
as an old family bible this is all
that I have time to write now so I will
close with my love to all from your affectionate
brother Dell

[4] hole where an indain chief called the Wild Cat cut his way out through the stone like fort with a knife this Wild Cat was a Chief that they had prisoner in the Florida war. he escaped from them 3 times and the last time he got away he was most starved and naked and they found his dead body in the woods a few days after there is an old gray headed man that lives near the barracks that tells us meny a rough old tale about that old fort it has been scaled twice by the indains and they were driven back boath times with an awfull slaughter I tell you there has been some tough fighting in this place there is the history of this place in the post office but it cant be got for love nor money as it is the only one round here it is near as large as an old family bible this is all that I have time to write now so I will close with my love to all from your affectionate brother Dell

rec'd Aug. 2 Jan. 20/64 A. H. Buffum.

St Augustine Florida January 3rd/64
Dear Brother
I now seat myself to write you a few lines it is some time since I got a letter from you but I suppose you dont have as much time to write as you used to but I get papers most evry from you so that makes up for it. It quite cold weather here now by spells it rains like the devil to day they have 2 companies D & F into tents the other day so we have got plenty of room now and we have bully old times here in the evenings New Years day I had a stavin old time I was on guard the niggars of Agustine had quite a time they had a large room and a table set and the old nigs got up and made speeches and then they had a free dinner for all the guard was all invited up and after dinner they had danceing they had quite a merry time of it.
I dont see your picture yet hope to soon I have got all the family but you

St. Augustine Florida January 3rd/64

Dear Brother

I now seat myself to write you a few lines it is some time since I got a letter from you but I suppose you dont have as much time to write as you used to but I get papers most evry ["mail" missing] from you so that makes up for it, it quite cold weather here not by spells it rains like the divel to day they have 2 companies D & F into tents the other day so we have got plenty of room now and we have bully old times here in the evenings New Years day I had a stavin old time I was on guard the niggers of Agustine had quite a time they had a large room and a table set and the old nigs got up and made speeches and then they had a free dinner for all the guard was all invited up and after dinner they had danceing they had quite a merry time of it. I dont see your picture yet hope to soon I have got all the family but you

Edgar this is the divels own hole there is every darn kind of athing to look after round a pickit post rebels and wildcats and marsh ponies and skunks and owls and nigger wenches there was a wildcat shot on pickit a few days ago the bugger came in to the tent and sot in a corner with glaring eyeballs looking at the guard untill they put 8 shots into him he was as large as a good sized bull dog the fellows that shot him belonged to Co. A and they brought him in and skun him and made 2 caps out of the skin he was a savage looking bugger I tell you he could take and chaw a mans leg up as easy as nothing, and here 3 days ago the rebels captured 20 of the 10th and killed 1 and captured 3 of our pioneers and wounded Leint Warker of Co. D of our Regt, our pioneers have to go 2 miles outside of our pickit to cut wood as there is none eny nearer but small brush so they have to go out there and the 10th and our

[2] Edgar this is the divels own hole there is every darn kind of athing to look after round a pickit post rebels and wildcats and marsh ponies and skunks and owls and nigger wenches there was a wildcat shot on pickit a few days ago the bugger came in to the tent and sat in a corner with glaring eyeballs looking at the guard untill they put 8 shots into him he was as large as a good sized bull dog the fellows that shot him belonged to Co A and they brought him in and skin him and made 2 caps out of the skin he was a savage looking bugger I tell you he could take and chaw a mans leg up as easy as nothing and here 3 days ago the rebels captured 20 of the 10th and killed 1 and captured 3 of our pioneers and wounded Liut Warker of Co D of our Regt our pioneers have to go 1 miles outside of our pickit to cut wood as there is none eny nearer but small brush so they have to go out there and the 10th and our

regt. take turnes going out as guard it takes 30 men for guard and the 10th happened to be on that day and they went out the same as usual and walked right into an ambush and they wounded the Leiut and killed one of the 10th and run off with 20 of the 10th and 3 of our pioneers and as they were mounted they were soon out of sight it was the bigest haul they ever made at this place since ~~since the war~~ the war commenced but the woodchopers did not have to go out untill lately they always cut inside the lines before till they cut the wood all down, and ever since that muss they have 100 men for guard and as soon as they get outside the pickit they put out skirmishers and skirmish way out to the place where they go to work so now the rebs cant surprise us so again we was out on Brigade Drill when the alarm was given and we went out on the double quick with our best cloaths on and did

[3] regt take turnes going out at guard it takes 30 men for guard and the 10th happened to be on that day and they went out the same as usual and walked right into an ambush and they wounded the Leiut and killed one of the 10th and run off with 20 of the 10th and 3 of our pioneers and as they were mounted they were soon out of sight it was the bigest haul they ever made at this place since the war commenced but the woodchopers did not have to go out untill lately they always cut inside the lines before till they cut the wood all down and ever since that muss they have 100 men for guard and as soon as they get outside the pickit they put out skirmishers and skirmish way out to the place where they go to work so now the rebs cant surprise us so again we was out on Brigade Drill when the alarm was given and we went out on the double quick with our best cloaths on and did

not get back untill 4 oclock in the afternoon if Chipman had stayed with the regt he would proberbly have got it instead of Leuit Warker as Chipman would have been 2nd Leuit of Co. D and as he went into the 5th Warker ~~the place~~ got promoted ~~to~~ 2nd Leuit of Co. D he was promoted since we came to Augustine he had not been back from his furlough only about a fortnight

I will send you a cross malle out of a piece of the old treasuary chest that the Spanish brought there money over in to pay there Soldiers off with I got a piece out of the Fort it is against orders to take any but one day I was in the fort and I took and cut a piece off and shoved it into my pocket and when I got time I made this cross of it I dont know as you would care enything about it so if you will give it to mother I think she would like it as it came from Florida and is so old that chest came here when this place was first built I dont know as I can think of enything more to write about now so I will close with my love to all, from your brother A. A. Buffum

from Dell to Edgar.

[4] not get back untill 4 Oclock in the afternoon if Chipman had stayed with the regt he would proberbly have got it instead of Leuit Warkr as Chipman would have been 2nd Leuit of Cd D and as he went into the 54 Warker ~~the place~~ got promoted to 2nd Leuit of Co D he was promoted since we came to Augustine he had not been back from his furlough only about a fortnight

I will send you a cross made out of a piece of the old treasury chest that the Spanish brought there money over in to pay there Soldiers off with I got a piece out of the Fort it is against orders to take any but one days I was in the fort and I took and cut a piece off and shoved it into my pocket and when I got time I made this cross of it I don't know as you would care enything about it so if you give it to mother I think she would like it as it came from Florida and is so old that chest came here when this place was first built I dont know as I can think of enything more to write about now so I will close with my love to all, from your brother A.A. Buffum

from Dell to Edgar

St. Augustine, Florida Feb 1st/64

ans. Feb. 27/64

(Rec^d Saturday Feb. 20. from Sergeant Cobb, Orderly of Co B 24th Regt.)

St Augustin Florida Feb 1st/64

Dear Brother

I recieved your letter last night was
very glad to here from you. it seemes
you dont think much of the reenlisting
business. I dont believe that the
expreshions you used in your letter were
your feelings. I think that mother
must have got you to write so to keep
me from reenlisting. because it dont
sound like you as you always wrote
different before. but you need not
worry eny more about it for I have
reenlisted. I wanted the furlough
most of enything but I wanted the
money to so I enlisted but the doctor
would not pass me on account of
my diarrhea. I was mad as the
divel but could not help myself as
he has got the whole say about it
so I will serve my time out and

St. Augustin Florida Feb 1st/64

Dear Brother

I recieved your letter last night was very glad to hear from you, it seemes you dont think much of the reenlisting buisness. I dont believe that the expreshions you used in your letter was your feelings. I think that Mother must have got you to write so to keep me from reenlisting. because it dont sound like you as you always wrote different before. but you need not worry eny more about it for I have reenlisted. I wanted the furlough most of enything but I wanted the money to so I enlisted but the doctor would not pass me on account of my diarrhea. I was mad as the divel but could not help myself as he has got the whole say about it so I will serve my time out and then come home for good. there was 54 men enlisted out of my company but there is some 6 or 7 say they wont go if I dont we all put our names down the last of Dec and in the first part of January but the mustering officier has not come down here yet and a good many of the boys have got tired of waiting for thire furlough and a great many of them say that now when he does come they will not get sworn in when we enlisted we expected to get the furlough right off now proberly they will not get home before march.

Feb 10th since writing the above I have been on a 3 days tramp and been on guard twice it is now about 12 o'clock and the sergeant of the guard has just turned in after calling me so I thought I would pass the time away by finishing your letter. when I left off writing before we were ordered to fall in for

[2] then come home for good. there was 54 men enlisted out of my company but there is some 6 or 7 say they wont go if I dont we all put our names down the last of Dec and in the first part of January but the mustering officier has not come down here yet and a good many of the boys have got tired of waiting for their furlough and a great many of them say that now when he does come they will not get sworn in when we enlisted we expected to get the furlough right off now proberly they will not get home before March.

Feb 10th since writing the above I have been on a 3 days tramp and been on guard duty twice it is now about 12 1/2 oclock and the sergeant of the guard has just turned in after calling me so I thought I would pass the time away by finishing your letter. when I left off writeing before we were ordered to fall in for

drill so I put the letter in my knapsack
and it has been there till now. we came
in from drill and went on dress
parade after we got back from dress
parade the cooks had orders to get 3 days
rations cooked by morning then we
knew that there was a tramp of them
the Lieut came over and told us we were
to start at daylight in the morning he
was detailed for guard and the Capt
was sick so we had to go without
an officier. so at 4 oclock we got up
and got breakfast and rations and formed
the line down in the plaza at daylight
and started 6 companies of our Regt and
14 of the teamsters mounted to act as
cavelry we marched till 1 oclock when
we went into camp it was about 15
miles out and about half the way we
had to march through water half way up
to our knees we layed round there
for 36 hours while the horsemen went

[3] drill so I put the letter in my knapsack and it has been there till now. we came in from drill and went on dress parade after we got back from dress parade the cooks had orders to get 3 days rations cooked before morning then we knew that there was a tramp up then the Leiut came over and told us we were to start at daylight in the morning he was detailed for guard and the Capt was sick so we had to go without an officier. so at 4 oclock we got up and got breakfast and rations and formed the line down on the plaza at daylight and started 6 companies of our regt and 14 of the teamsters mounted to act as caverly we marched till 1 oclock when we went into camp it was about 15 miles out and about half the way we had to march through water half wa[y] up to our knees we layed round there for 36 hours while the horsemen went

out and drove in the cattle as that was
what we went after they brought in 7 rebs
and drove in 73 head of cattle and we
started back and got home just at dark
after being out 2 days and a half it was
cold as the divel all the time we were
out and we had nothing but a rubber blanket
as we did not want to take anything more
and in the night time we were as cold as
the divel and we were not allowed to build
a fire for fear the rebs would come down
on us, I tell you I was glad when we got
home. to day the boys signed a new muster
roll they got 49 men of Co. B they are going
to start for Hilton Head the day after to morrow
and I expect us fellows that dont go will
get guard duty like the dibel till the V.s
get back from there furlough all right in
about 7 months I want to have a good long
furlough and I hope to shall be home
and see you all good and hearty

this from your brother Deck

in haste

[4] out and drove in the cattle as that was what we went after they brought in 7 rebs and drove in 73 head of cattle and we started back and got home just at dark after being out 2 days and a half it was cold as the divel all the time we were out and we had nothing but a rubber blanket as we did not want to tote enything more and in thc night time we were as cold as the divel and we were not allowed to build a fire for fear the rebs would come down on us, I tell you I was glad when we got home, to day the boys singed our new muster roll they got 49 men of Co B they are going to start for Hilton Head the day after tomorro and expect us fellers that dont go will get guard duty like the divel till the VVs [?] get back from there furlough all right in about 7 months I will take a good long furlough and I hope I shall be home and see you all good and hearty, this from your brother Dell, in haste

Jacksonville, Florida, March 18th 1864

Adelbert.

Answered 13th April 1864.

Jacksonville March 18th 1864

Dear Brother

I now take time to write you a few lines to let you know that I am well. I have recieved 3 letters from you since I wrote last I would have wrote before but we have about as much as we can do here now. our regt. is doing the provost here and as we have so few men it brings us on every other day and sometimes 48 hours at a time. Col Osborn is comander of the post. we are in camp in our old tents that we had on Morris Island in an old squar that was burned down near the centre of the town I suppose we shall have to move camp when the regt. gets back. the day after the Boys went home I was on pickit at Augustine and when I got relieved the next moring I found everything packed up ready to start

Endorsed in CEB's handwriting "Adelbert. Answered 13" April 1864."

Jacksonville March 18th 1864

Dear Brother

I now take time to write you a few lines to let you know that I am well. I have recieved 3 letters from you since I wrote last I would have wrote before but we have about as much as we can do here now, our regt is doing the provost here and as we have so few men it brings us on every other day and sometimes 48 hours at a time. Col Osborn is comander of the post. we are in camp in our old tents that we had on Morris Island in an old squar that was burned down near the centre of the town I suppose we shall have to move camp when the regt gets back. the day after the Boys went home I was on pickit at Augustine and when I got relieved the next morning I found everything packed up ready to start

off for Jacksonville. we got here just about
dark when Col Osborn took comand of Post
and we was put on provo and have been
doing that duty ever since we relieved
the 55 regt. colored the first night we were
here I saw a nigger hung the same day
there was 2 more hung up at Camp Finegan
for forcing a white women they belonged
to Co. H 55th regt. and a little while ago
I saw a nig shot for mutiny I tell you
they shove the nigs right through here if
they dont tread up. Chipman is here
he has been over to our camp 2 or 3 time
there is about 12 officers in the nigger
regts. here that used to belong to our regt.
you say that you saw the name of Major
Bogle in the list of killed he is not
killed but wounded and in the rebels
hands and that is worse I suppose.
after we had been here 2 or 3 days the
troops began to come in and you should
heard them swear and curse Seymour

[2] off for Jacksonville. we got here just about dark when Col Osborn took comand of Post and we was put on provo and have been doing that duty ever since we relieved the 45 regt colered the first night we were here I saw a nigger hung the same day there was 2 more hung up at Camp Finegan for forceing a white woman they belonged to Co H 55th regt and a little while ago I saw a nig shot for mutinny I tell you they shove the nigs right through here if they dont tread up. Chimpman [?] is here he has been over to our camp 2 or 3 times there is about 12 officers in the nigger regts here that used to belong to our regt you say that you saw the name of Major Bogle in the list of killed he is not killed but wounded and in the rebels hands and that is worse I suppose. after we had been here 2 or 3 days the troops began to come in and you should heard them swear and curse Seymour

for takeing them up to so many rebel
he might know he would get licked
and they did get licked like the devil
but they expect to try it again soon.
I was quite surprised to here that Father
had joined the free masons but I dont
suppose it will do him eny hurt I
am sorry for Albert Crockett for I think
he will have a hard time of it I guess I
shall get home before he does now
how is Tedd now does he want to
be a soldier as bad as he did tell him
if he will take my advice he will
keep out of the army altogether he cant
think what it is like untill he gets
out here and trys it so say that I tell
him he had better stay at home
tell Father that I recieved his letter
with the money in it all right and I
will answer it as soon as I can get
time. I am going to write to Jonathan
as soon as I can. was there a fellow by

[3] for takeng them up to so many rebels he might know he would get licked and they did get licked like the divel but they expect to try it again soon.

I was quite surprised to here that Father had joined the free masons but I dont suppose it will do him eny hurt I am sorry for Albert Crockett for I think he will have a hard time of it I guess I shall get home before he does now

how is Tede now does he want to be a soldier as bad as he did tell him if he will take my advice he will keep out of the army altogether he cant think what it is like untill he gets out here and trys it so say that I tell him he had better stay at home

tell Father that I recieved his letter with the money in it all right and will answer it as soon as I can get time. I am going to write to Jonathan soon as I can. was there a fellow by the name of Ned Carthy at the store yet I sent a piece of the oldest house in Augustine by him he is a Corporal in my company he said he would leave it at the store his folks live round there somewhere I told a good many of the boys to call round at the store and to send there folks there to trade.

so you are 21 years old I pass for most that that is not more than 2 in the rest that knows my age I pass for 20 last August just 2 years older than I am I told them I was 18 when I enlisted so I let it run on I could not have enlisted if I had not been 18 you know now sometimes I almost forget how old I realy am I vow. Edgar I dont see that picture of yours that you promised to send me I hope you will send it soon, tell Teddy to send me one of his with his coat and vest on but be sure and send me yours I have got all the family but you. give my love to all the folks

from your Brother Dell,

[4] the name of Ned Carthy at the store yet I sent a piece of the oldest house in Augustine by him he is a Corporal in my company he said he would leave it at the store his folks live round there somewhere I told a good many of the boys to call round at the store and to send there folks there to trade.

so you are 21 years old I pass for most that that is not more than 2 in the regt that knows my age I pass for 20 last August just 2 years older than I am I told them I was 18 when I enlisted so I let it run on I could not have enlisted if I had not been 18 you know that sometimes I almost forget how old I realy am I vow. Edgar I dont see that picture of yours that you promised to send me I hope you will send it soon, tell Tede to send me one of his with his coat and vest on but be sure and send me yours I have got all the family but you, give my love to all the folks

from your Brother Dell.

IN THE FIELD, MAY 8, 1864

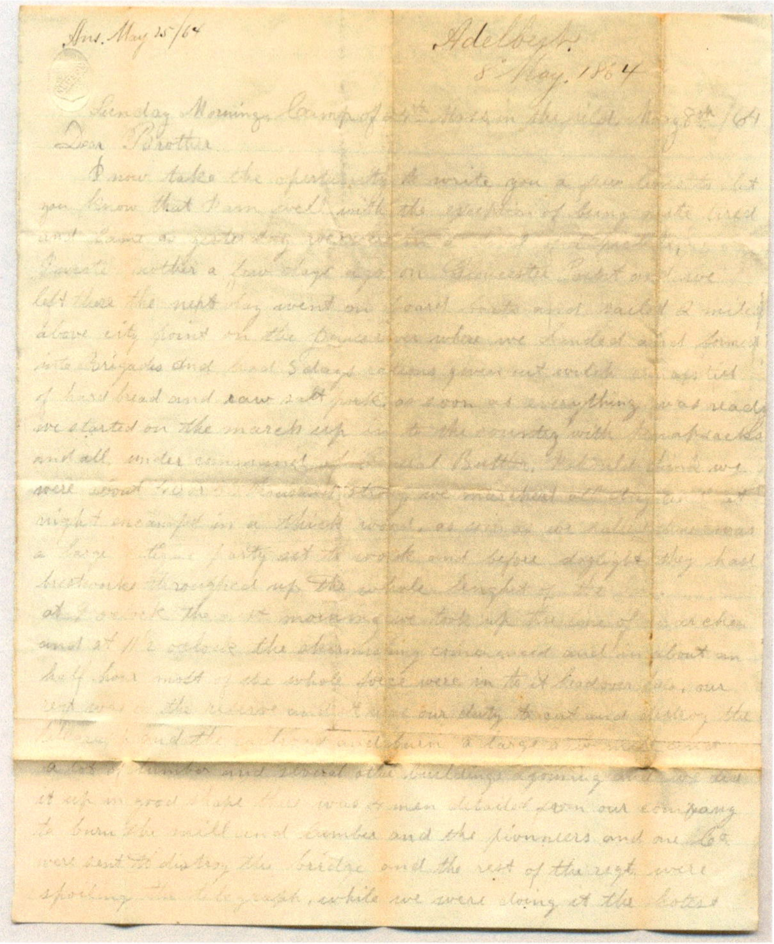

Ans. May 25/64

Adelbert.
8th May. 1864

Sunday Morning, Camp of [illegible] Mass in the field May 8th /64
Dear Brother

I now take the opertunity to write you a few lines to let you know that I am well with the exception of being quite tired and lame [illegible] yesterday [illegible]
I wrote mother a few days ago on Gloucester Point and we left there the next day went on board boats and sailed 2 miles above city point on the James river where we landed and formed into Brigades and had 5 days rations given out which consisted of hard bread and raw salt pork. as soon as everything was ready we started on the march up in to the country with knapsacks and all under command of General Butler. [illegible] we were about [illegible] thousand strong we marched all day [illegible] at night encamped in a thick wood. as soon as we halted there was a large fatigue party set to work and before daylight they had breastworks throughed up the whole length of the [illegible]
at [illegible] o clock the next morning we took up the line of march and at 11½ o clock the skirmishing commenced and in about an half hour most of the whole force were in to it [illegible] our regt was in the reserve and it was our duty to cut and destroy the [illegible] and the [illegible] and burn a large [illegible]
a lot of lumber and several other buildings adjoining and we did it up in good shape there was [illegible] men detailed from our company to burn the mill and lumber and the pionneers and one Co. were sent to destroy the bridge and the rest of the regt. were spoiling the telegraph. while we were doing it the hottest

Endorsed in CEB's handwriting "Adelbert. 8" May, 1864. Ans. May 25/64."

Sunday Morning. Camp of 24th Mass. in the field.
May 8th/64.

Dear Brother

I now take the opertunity to write you a few lines to let you know that I am well with the exception of being quite tired and lame as yesterday we were in a divel of a pickle. I wrote mother a few days ago on Gloucester Point and we left there the next day went on board boats and sailed 1 miles above city point on the James river where we divided and formed into Brigades and had 3 days rations given out witch concisted of heard bread and raw salt pork. as soon as everything was ready we started on the march up in to the country with knapsacks and all under command of General Buttler. I should think we were about 40 or 50 thousand strong, we marched all day and at night encamped in a thick wood. as soon as we halted there was a large fatigue party set to work and before daylight they had brestworks throughed up the whole lenghet of the line. at 9 oclock the next morning we took up the line of marche and at 11 1/2 oclock the skirmishing commenced and in an about an half hour most of the whole force were in to it head over heels. our regt was in the reserve and it was our duty to cut and destroy the telegraph and the railroad and burn a large saw mill and a lot of lumber and several other buildings ajoining and we did it up in good shape there was 4 men detailed from our company to burn the mill and lumber and the pionniers and one Co were sent to distroy the bridge and the rest of the regt were spoiling the telegraph, while we were doing it the hotest

firing was going on that ever I heard in my life it was an awful hot fight, we had no more than got through with our work than the order came to retreat and we took our axes in our hands and walked back to where we encamped the night before about as quick as we could, our regt came out without losing a man but some of the regts got cut up [illegible] bad, the 8 Maine [illegible]
Chuck Collins [illegible] I saw them at Gloucester Point just before we came away but have not heard from them since. I should think our loss must be very heavy we have not heard the particulars yet and [illegible] tell anywheres near how much it is, the fight was at the Virginia Central Railroad where the rebels take all their supplies [illegible] to Richmond [illegible] it will [illegible] understand we are to try it [illegible] again [illegible] I hope we shall meet with better success than we did this time. we have Generals Butler and Gillmore, and Smith and Brooks, here with us
we have had quite rough times since we joined the veterans nothing to eat but fat raw salt pork and hard bread but for all that I feel quite smart this climate agrees with me better than farther south I think for I am getting quite strong now but there is one thing I am [illegible] thank all that I am not a veteran I have not time to write more now but will write again as soon as anything new turns up [illegible] give my love to the folks and to that girl of yours up in New Hampshire I remain your affectionate brother,

From your brother Dell

P.S. please write soon and send that picture of yours along. Dell

[2] fireing was goning on that ever I heard it my life it was an awfull hot fight, we had no more than got through with our work than the order came to retret and we took our ass in our hands and walked back to where we encamped the night before about as quick as we could, our regt came out without loseing a man but some of the regts got cut up very bad, the 8 Maine caught it pretty heavy that is the regt that Charlie Davis and Chuckie Elliot from Melrose belongs to. I saw them at Gloucester Point just before we come away but have not heard from them since. I should think our loss must be very heavy we have not heard the particulars yet and cant tell eny wheres near how much it is, the fight was at the Virginia Central Railroad where the rebels take all their suplies over to Richmond so they had it well guarded. I understand we are to try it over again in a day or two when we do I hope we shall meet with better success than we did this time. we have Generals Buttler and Gillmore and Smith and Brooks, here with us

we have had quite rought times since we joined the veterans nothing but fat raw salt Pork and hard bread but for all that I feel quite smart this climate agrees with me better than farther south I think for I am geting quite strong now but there is one thing I am quite thanksfull that I am not a veteran I have not time to write more now but will write again as soon as enything new turns up so give my love to all the folks and to that girl of yourn up in New Hampshire I remain your affectionate brother,

from your brother Dell.

PS. plese write soon and send that picture of yours along,

Dell.

In the Field, May 11, 1864

Ans. May 25/64

Camp of 24th Mass [illegible] in the field May [illegible] 1864

[illegible]

[illegible] you [illegible] this month since [illegible] we have been on the move all the time [illegible] was the first nights sleep we have had for five days. [illegible] had finished writing that letter we had orders to fall in [illegible] marched back about [illegible] and pitched camp and got a little [illegible] and the whole regt was ordered [illegible] building a fort we [illegible] there until [illegible] had orders to [illegible] we came onto camp and started [illegible] about 2 miles [illegible] went on picket [illegible] just about [illegible] the orders was [illegible] keep awake [illegible] very careful not [illegible] that night. we were kept [illegible] without [illegible] of sleep and [illegible] night [illegible] relieved and came down to camp [illegible] got a [illegible] nights [illegible] but I think we were lucky again [illegible] being on picket for if we had not been we should have had to go into the fight as they were fighting like the devil all the time we were on the [illegible] we could hear the rebels [illegible] in the woods [illegible] the road we could hear them hollering all night [illegible]

[Endorsed in CEB's handwriting "Ans. May 25/64"]

Camp of 24th Mass Volls. in the field May 11th 1864

Dear Brother

I wrote you last on the 8th of this month since when we have been on the move all the time last night was the first nights sleep we have had for five days.

As soon as I had finished writeing that letter we had orders to fall in and we marched back about 1 mile and pitched camp and got a little grub and the whole regt was ordered out on fatuigue building a fort we worked there untill 4 oclock when we had orderes to knock off and go pickit, so we came into camp and started up to the front about 2 miles and went on pickit we got posted just about dark the orders was for all hands to keep awaik to be carfull not to close our eyes that night, we were kept on 2 days and 2 nights without a wink of sleep and last night we were relieved and came down to camp and got a bully nights sleep. but I think we were lucky again in being on pickit for if we had not been on we should have had to go into the fight as they were fighting like the divel all the time we were on the first night we were we could here the rebels round in the woods falling trees blocking the road and we could here them hollering all night, a

as soon as day light the troops began to advance out [illegible]
Ben Buttler and Gillmore and all of there staff [illegible]
passed my post and a lot of nigger cavelry so I had
had a good chance to see all that was going on
[illegible] a little ways from
my post and took out there maps and began to look
them over when they was fired upon by the rebels and old
[illegible] took off 2 or 3 picket posts and deployed them out
in the woods and drove the rebs away and then finished
what he was about it seemed strange that the rebs did
not fire before for they were right in sight of our picket
and so [illegible] the rest had been going right by them and
they did not fire a shot.
[illegible] after the troops went out the [illegible]
and brisk firing was heard they were fighting most
all day but along towards the evening stopt but at about
midnight they went at it again for about a couple of hours
and the next day they were at it again at about 9 oclock
and kept at it all day and just before we were relieved
the fighting stopt and all has been quiet since
the troops that we was fighting come from Richmond
to drive us off this point at there only chance but
they could not come it. we heard yesterday that
Grant was giving the rebs hell and that the
rebel General Longstreet was killed and Lee and
Hill wounded the news was all over the whole
forces so I suppose it must be thus. we had
reinforcements of about 15 thousand from Grant

[2] as soon as daylight the troops began to advance out Ben Buttler and Gillmere and all of there staff passed my post and a lot of nigger cavelery so I had a good chance to see all that was going on General Buttler and Gillmore road out a little way from my post and took out ther maps and began to look them over when they were fired upon by the rebels and old Ben took off 2 or 3 pickit posts and deployed them out in the woods and drove the rebs away and then finished what he was about it seemed strange that the rebs did not fire before for they were right in sight of our pickit and regt after regt had been going right by them and they did not fire a shot.

about an hour after the troops went out the fight commenced and shuch fireing I never heard. they were fighting most all day but along night the fireing stopt but about midnight they went at it again for about a couple of hours and the next day they were into it again at about 9 oclock and kept at it all day and just before we was relieved the fighting stopt and all has been quiet since

the troops that we was fighting came from Richmond to drive us off this point as there only chance but they could not come it, we heard yesterday that Grant was giveing the rebels h—l and that the rebel General Longstreet was killed and Lee and Hill wounded the news was all over the whole forces so I suppose it must be true. we had reinforcements of about 10 thousand from Grant

last night I suppose there will be plenty of [illegible]
fighting [illegible] us [illegible] a day or two we dont
know what minnute we will be ordered out
for we are [illegible] not [illegible] have [illegible] round our tents
[illegible]
our boys are hard up for tobacco here as there is not
a sutler on the place I have not got a chew and
I had to borrow 2 postage stamps to write you Letter
I wish you would ask father if he cant send
me a little tobacco by mail for that is the only
way we can get [illegible] and tell him to send some
postage stamps
all of the stuff that we packed up in [illegible]
[illegible]
that it was [illegible] by one of our ironclads
off Fort [illegible] and sunk immediately
all the clothing and little nicknacks that the boys
packed up got lost and all the company books
and all the trunks of the officers most of the officers
lost everything except what they stand in all
went to the bottom I packed up everything I had
but my blouse and one pair of pants and 2 shirts and
2 pair of socks and the testament that mother gave
me in [illegible] everything else I had is at the
bottom of the [illegible] but I dont care much
I had rather lose them than [illegible] them round
I had to [illegible] away my blanket coming from
the landing up here and have been without [illegible]

[3] last night I suppose there will be plenty more fighting for us in a day or two we dont know what minnit we will be ordered out for we are ordered not to leave from round our tents so I thought I would write you while I had a chance our boys are hard up for tobacco here as there is not a sutler on the place I have not got a chaw and I had to borrow 2 postage stamps to write you lately I wish you would ask father if he cant send me a little tobacco by mail for that is the only way we can get here and tell him to send some postage stamps.

all of the stuff that we picked up on Gloucester Point to send to Fortress Monroe is lost the Steamer that it was on got run in to by one of our ironclads off Fort Monroe and sunk immediatly all the cloathing and little nicknacks that the boys packed up got lost and all the company books and all the trunks of the officies most of the officies lost evering thing except what they stand in all went to the bottom I packed up everingthing I had but my blouce and one pair of pants and 2 shirts and 2 pair of socks and the testement that Mother gave me in redeville everingthing else I had it at the bottom of the hamton roads but I dont care much I had rather lose them than to lug them round I had to through away my blanket comeing from the landing up here and have been without ever

since we can't draw nor buy anything here
I don't know how long this thing is going to last
nor I don't care if we do what is expected round here
[illegible] see that Grant dont want to take Richmond
[illegible] surround them [illegible] about Vicksburg and bag the
whole rebel army but he will have [illegible] to
do that and there will have to be some awful [illegible]
[illegible] before [illegible] but I think the war is
going to be settled [illegible]

I have got to go down and wash my clothes now
while I have [illegible] will write
again as soon as [illegible]

[illegible]

affec[illegible]

[illegible]
Direct [illegible]
Co [illegible] Regt [illegible] Monroe

[illegible]
11 Aug 1864

[4] since we cant draw nor buy ennything here I dont know how long this thing is going to last nor I dont care if we do what is expected round here the talk is here that Grant dont want to take Richmond right off he is going to come [unreadable] and surround it as he did Vicksburg and bag the whole rebel army but he will have enough to do to do that and there will have to be some awfull hard fighting before he could do it but I think the war is going to be settled this year.

I have got to go down and wash my cloathes now while I have time so I must close but will write again as soon as enything new turns up

give my love to all the folks from your affectionate Brother

P.S. direct to Corporal A.A. Buffum

Co B 24th Regt Mass Volls Via Fortress Monroe

from

B Buffum [sic]

In the Field, May 19, 1864

Rec May 27/8
Answered May, 25, 1864
Adelbert

Head Quarters in [illegible] field May 19th 186[illegible]

Dear Brother

I now have time to write you a few lines to let you know that I am [illegible] kicking we have seen any quantity of fighting since I wrote last. we have been on a tramp 8 days with nothing but rubber blankets to cover us from the rain and cold we left overcoats and everything else in camp. we have been within [illegible] miles of Richmond I got a piece of the [illegible] mile [illegible] and a piece of the telegraph wire in my pocket. we [illegible] [illegible] 8 days and [illegible] [illegible] [illegible] [illegible] Mr Reb [illegible] [illegible] up and got [illegible] we had to make lively on the [illegible] of the 7 the rebs kept still as mice but in the morning early the firing commenced along our whole line it was so foggy we could not see [illegible] before us and the rebels took advantage of it and crept close up on to us before we knew it but after a while the fog cleared away and the sun came out bright and hot and the firing increased rapedly, at last the rebs drove back our [illegible] and our [illegible] [illegible] not [illegible] [illegible] [illegible] [illegible] [illegible] [illegible] [illegible] had orders to retreat and the way we had to [illegible] over the ground was not slow untill we had gained a position on a hill when we turned on them and the way them rebels went scrambling back was laughable

[Endorsed in CEB's handwriting "Rec'd May 24

Answered May, 25, 1864.

Adelbert."]

Head Quarters in the field May 19th 1864

Dear Brother

I now have time to write you a few lines to let you know that I am alive and kicking we have seen eny quanterty of fighting since I wrote last. we have been on a tramp 8 days with nothing but rubber blankets to cover us from the rain and cold we left overcoats and everthing else in camp we have been within 10 miles of Richmond I got a piece of the 10 mile stone and a piece of the telegraph were in my pocket. we fight them 8 days and licked them every day but on the 9 day Mr Reb told us to get up and get and we had to quite lively on the night of the 8 the rebs kept still as mice but in the morning early the fireing commenced along our whole line it was so fogy we could not see 4 feet before us and the rebels took advantage of it and crept close up on to us before we knew it but after a while the fog cleared away and the sun came out bright and hot and the fireing incresed rapedly. at laast the rebs drove back our center and our regt did not know it untill the rebs were right on top of us then we had orders to retreat and the way we had to kite over the ground was not slow untill we had gained a position on a hill when we turned on them and the way them rebels went scrambling back was laughable we kept up a hot fire till we drove them back out of sight when we stoped and the General commanding our brigade formed us in line and went through dress parade right in sight of the rebels he said there was nothing better in the world to steady the mens nerves. I heard the General tell our Colonel that he thought our reg't was all captured he said if we had been 10 minnits later or had not come out as we had we should have all been taken we commenced the retreat at about 12 oclock and got to camp at 9 at night our reg't was the rear guard coming back our reg't lost 3 officers killed and one wounded and 53 men killed and wounded and 8 missing yesterday we were on picket again the rebels have followed [illegible]

day yesterday and untill 10 oclock today our reg't was relieved last night the rebels are trying to siege us out of this place every reg't has got rifle pits in front of their camp so we have work enough all the time I have not time to write more now but will write again as soon as I have time I have wrote 3 times to you since I have received no letter from you or any of the other folks at home

[illegible] my love to all the folks

[illegible] from your Brother [illegible]

in haste

[2] we kept up a hot fire till we drove them back out of sight when we stoped and the General commanding our brigade formed us in line and went through dress parade right in sight of the rebs he said there was nothing better in the world to steady the mens nerves. I heard the General tell our Colonel that he thought our regt was all captured he said if we had been 10 minnits later or had not come just as we had we should have all been taken we commenced the retreat at about 10 oclock and got to camp at 9 at night our regt was the rear guard comeing back our regt lost 3 officiers killed and one wounded and 53 men killed and wounded and 8 missing

yesterday we were on pickit again the rebels have followed us clear up to our fortifycations they were fighting all day yesterday and untill 10 oclock today our regt were relieved last night. the rebels are trying to siege us out of this place every regt has got rifle pits in front of there camps so we have work enough all the time I have not time to write more now but will write again as soon as I have time I have wrote 3 times to you since I have recieved no letter from you or eny of the other folks at home

give my love to all the folks

this from your Brother Dell

in haste

In camp, June 1, 1864

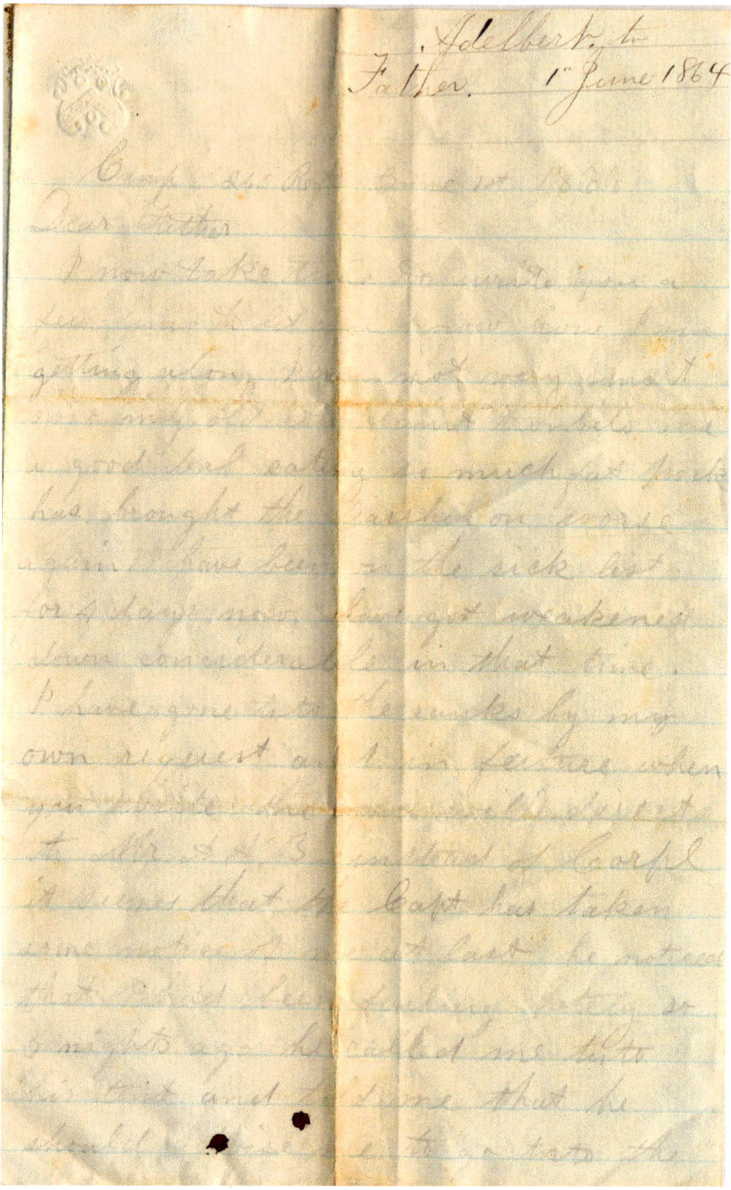
Adelbert to
Father. 1st June 1864

Camp 24th Reg[illegible] June 1st 1864
Dear Father
I now take th[illegible] to write you a
few [illegible] to let you know how I am
getting along. I [illegible] not very smart
[illegible] old [illegible] bowels and
a good deal eating so much fat pork
has brought the [illegible] on worse
again. I have been on the sick list
for 4 days now. [illegible] got weakened
down considerable in that time.
I have gone into the ranks by my
own request and [illegible] in future when
you write [illegible] will direct
to Mr A H B instead of Corpl
It seems that the Capt has taken
some notice [illegible] at last. he noticed
that [illegible] been [illegible] lately so
3 nights ago he called me into
his tent and told me that he
should [illegible] to go into the

[Endorsed in CEB's handwriting "Adelbert. to Father. 1 June 1864."]

Camp of 24th Regt June 1st 1864.

Dear Father

I now take time to write you a few lines to let you know how I am getting along I am not very smart now my old complaint troubels me a good deal eating so much fat pork has brought the diarrhea on worse again. I have been on the sick list for 4 days now. I have got weakened down considerable in that time. I have gone into the ranks by my own request and in future when you write me you will direct to Mr. A.A. B. instead of Corpl it seems that the Capt has taken some notice of me at last he noticed that I had been failing lately so 3 nights ago he called me into his tent and told me that he should like me to go into the ranks he said I had been sick a good deal lately and if I went into the

ranks he said I had been sick
a good deal lately and if I went
into the ranks it would be a good
deal better for me as there would not
be near as much expected nor asked
of me, well I told him I was willing
and he said all right that the order
would be made out in the form
of returned to the ranks at his own
request, he said it was no disgrace
nor dishonor but the best thing I could
do, well the next day I went to him
and told him that I would like
my discharge or to be sent off
somewhere so I would stand a
chance to get well or get better.
the capt said he would see the surgeon
and see what he could do for me
and that is the last I have heard of
it I dont suppose I could get sent off
to any hospital for there are all full
of wounded now but I dont think
I can get much better so long

[2] ranks it would be a good deal better for me as there would not be near as much expected nor asked of me. well I told him I was willing and he said all right that the order would be made out in the formed of returned to ranks at his own request. he said it was no disgrace nor disoner but the best thing I could do. well the next day I went to him and told him that I would like my discharge or to be sent off somewhere so I would stand a chance to get well or get better. the Capt said he would see the surgeon and see what he could do for me and that is the last I have heard of it I dont suppose I could get sent off to eny hospital for there are all full of wounded now but I wont think I can get much better as long

as I stay here living on the fat
pork all the time, I dont do anything
now but lay round in the shelter
tent in the hot sun the doctor
excused me from all duty now for
a spell. quite a number of the boys
are comming down sick. we have had
duty on picket every third night
for 24 hours without a wink of sleep
the rebels within speaking distance
all the time on the outposts I have
been on duty all the time untill the
last 4 days. as soon as the regt comes
off picket they have to go on fatuigue
and so they keep them a humming
all the time. there has been a good
deal of shelling here lately. a great
many of the shells come over into our
camps but one good thing they have
stoped picket fireing here we dont
hear a musket shot from morning
till night but they make it up
with artillery fireing for several.

[3] as I stay here liveing on the fat pork all the time. I don't do enything now but lay round in the shelter tent in the hot sun the doctor excuses me from all duty now for a spell. quite a number of the boys are comeing down sick we have hard duty on pickit every third night for 24 hours without a wink of sleep the rebels within speaking distance all the time of the outposts I have been on duty all the time untill the last 4 days. as soon as the regt comes off pickit they have to go on fatuigue and so they keep them a huming all the time. there has been a good deal of shelling here lately a good many of the shells come over into our camps but one good thing they have stopped pickit fireing here we dont hear a musket shot from morning toll night but they make it up with artillery fireing for several

past there has [illegible] heavy firing
in the direction of [illegible] they have
been thumping away there awful
we expect to hear good news from
Grant in a few days
the boys are having boxes from home now
they come in 5 days so I thought I would
ask you to send me one you can
send [illegible] now and it will
not spoil it comes so quick, I wish you
would send me a [illegible] little tobacco [illegible]
and a piece of cheese mild cheese is the
kind I like a can of milk, a B. Sausage
a few nuts and anything else you see fit
[illegible] you can put a couple of [illegible]
of matches in a corner of the box. tell
mother she can put in a little cake
if she has a mind to. now if you fit
the box right up and send it along I
will get it in about a week from now
the mail comes here every night now
the boys all get their letters in 2 days
after they are mailed I received your
postage stamps and tobacco all safe [illegible] told
you in my last letter. some of the boys had a
pound of tobacco in the mail last night with
16 cents in postage on it rather a clear way
to send it. as I have [illegible] the [illegible]
of the page I must close with my love to all
[illegible] [illegible] [illegible]

[4] past there has been very heavy fireing in the direction of Richmond they have been thumping away there awfully we expect to hear good news from Grant in a few days

the boys are haveing boxes from home now they come in 5 days so I thought I would ask you to send me one you can send most anything now and it will not spoil it comes so quick. I wish you would send me a box a little tobacco and a piece of chese mild chese is the kind I like a can of milk a B. Saugage a few nuts and enything els you see fit to put in you can [illegible] couple of bunches of matches in a cornor of the Box. tell Mother she can put in a little cake if she has a mind to. now if you fit the box right up and send it along I will get it in about a weeak from now the mail comes here every night now the boys all get thir letters in 2 days after they are mailed I recieved your postage stamps and tobacco all safe as I told you in my last letter, some of the boys had a pound of tobacco in the mail last night with 96 cents in postage on it. rather a dear way to send it. as I have got down to the bottom of the page I must close with my love to all the folks at home I remain your affectionate Son

Deep Bottom Va July 24th, 1864

Adelbert.

Deep Bottom Va July 24th /64

Dear Brother

I recieved your letter with your picture yesterday and was very glad to get the picture I think it is a very good one. I am very glad Father would not let Theodore go off as I dont think the army a fit place for him as he would have found out if he had come. I hope he will know enough now to stay at home and do the best he can to help father.

I shall be home in the course of 2 months I suppose. I dont think there is eny sight for me to get home before my time is out Captain Gardner had a letter from General Dale concerning me but he did not tell me about it or say a word to me about a furlough or discharge or enything of the kind. the buggers if they find out enything

Deep Bottom Va July 24th/64

Dear Brother

I recieved your letter with your picture yesterday and was very glad to get the picture I think it is a very good one. I was very glad Father would not let Theodore go off as I dont think the army a fit place for him as he would have found out if he had come. I hope he will know enough now to stay at home and do the best he can to help father.

I shall be home in the course of 2 months I suppose. I dont think there is eny sight for me to get home before my time is out. Captain Gardner had a letter from General Dale concerning me but he did not tell me about it or say a word to me about a furlough or discharge or enything of the kind. the buggers, if they find out enything

they keep it to themselves.
I see by your letter that Dr Dale wrote to Buttler as well as the Capt so I suppose he has got it before this time and if he was a mind to he could get me discharged or if Dr Dale should demand my discharge they would be obliged to give it. he can tell how much I have sick by looking at the monthly reports of our the 24 Regt. wich are at the State House in Boston.
I suppose you could look at them if you did you would see that I have been sick with the diarhea for a year. that is if the reports are made out right I cant find out enything in this regt if I ask enything about thoes papers they tell me they dont know enything about it. the fact is the Dr dont want to discharge enyone out of the regt. he wants to have the name of having the healthiest of eny Mass Regt.
I wrote to Father a while ago and

[2] they keep it to themselves

I see by your letter that Dr Dale wrote to Buttler, as well as the Capt so I suppose he has got it before this time and if he was a mind to he could get me discharged or if Dr Dale should demand my discharge they would be obliged to give it. he can tell how much I have sick by looking at the monthly reports of ~~our~~ the 24 Regt wicth are at the State House in Boston

I suppose you could look at them if you did you would see that I have been sick with the diarrhea for a year that is if the reports are made out right I cant find out enything in this regt if I ask enything about thoes papers they tell me they dont know enything about it the fact is the Dr dont want to discharge enyone of the regt he wants to have the name of haveing the healthest of eny Mass Regt

I wrote to Father a while ago and

sent for a box but haive not got
eny answer from it yet I have
been waiting for an answer before I
write to him, they need not be affraid
to write thinking I am on the way home
for there is no danger of my comeing untill
my time is out.
Leiut Col Hooper of our regt. was taken prisoner
about a hour ago he was in commaind
of the pickit ~~kt~~ line and rode out a
little way from our vitette and was
fired on and taken prisoner I dont know
wether he was wounded or not
we have had reinforcements of a divistion
of the 19 army corps so I suppose the
duty will be a little easyer I am geting
better now I think I will go on duty in
a few days I'm darned if they can kill
me unless I am shot, we had one man
die a short time ago of the diarrhea
he was not sick more than 10 days
another that was wounded the 17 of June
died 3 more wounded ones home on

[3] sent for a box but have not got eny answer from it yet I have been waiteing for an answer before I write to him. they need not be affraid to write thinking I am on the way home for there is no danger of my comeing until my time is out.

Leiut Col Hooper of our regt was taken prisoner about a hour ago he was in command of the picket ~~let~~ line and rode out a little way from our vitette and was fired on and taken prisoner I dont know wether he was wounded or not we have had reinforcements of a divistion of the 19 army corps so I suppose the duty will be a little easyer I am geting better now I think I will go on duty in a few days Im darned if they can kill me unless I am shot. we had one man die a short time ago of the diarrhea he was not sick more than 10 days another that was wounded the 17 of June died 3 more wounded ones home on

furlough, so I think I have been
lucky to stand it as long as I have
I think I must have a good constitution
or I would have kicked the bucket
long ago
I wish you would get Theodore to have
his picture taken and send to me I
have got one of him that was taken
when he wore a jacket I should like
one of him now he must look so
different make him send me one
you can I know,
the cooks are calling me to help them
peel potatoes for the supe so I must
close

from your brother Dell

in haste

[4] furlough so I think I have been lucky to stand it as long as I have I think I must have a good constitution or I would have kicked the bucket long ago

I wish you would get Theodore to have his picture taken and send to me I have got one of him that was taken when he wore a jacket I should like one of him now he must look so different make him send me one you can I know.

the cooks are calling me to help them peel potatoes for the supe so I must close

from your brother Dell

in haste

Vitette=vidette: a mounted sentry positioned beyond an army's outposts to observe the movements of the enemy.

Made in the USA
Las Vegas, NV
03 November 2026

57425295R00112